Yearbook of the Artificial

Yearbook of the Artificial Volume 5

Edited by
Massimo Negrotti

In the different areas of the artificial – defined as the attempt to reproduce natural objects or processes by means of the available technology – researchers and designers often work within their disciplines with little or no information about research being carried out in other areas. Bioengineers, for instance, know little about the problems of Artificial Intelligence researchers, while roboticians often neglect the efforts of medical engineering, etc. The Yearbook of the Artificial discusses current theories, projects and models from a wide range of areas without going into too many technical details. The main assumption is that there are common logical and methodological problems for all disciplines that are concerned with the reproduction of natural objects and processes. The Yearbook provides a forum for all of them, and covers areas such as bioengineering, robotics, A.I., artificial neural networks, artificial life, multimedia, the history of technology, communications, art and music.

In submitting a paper to the *Yearbook of the Artificial*, you grant us permission to publish that paper after a peer review. We ask that your paper be between 20 000 and 24000 characters, and pertinent to the topics mentioned above. We also require an abstract of no more than 150 words, including a list of keywords. Papers should be submitted in triplicate, typewritten, double-spaced, on one side of the paper. Please include a copy on a floppy disk in a commonly used word processor format.

Only original papers in English should be sent to:

Dr. Domenico Donnanno
dodo@soc.uniurb.it
IMES-LCA
Via Saffi, 15
University of Urbino
I-61029 Urbino (Pu) Italy

Yearbook of the Artificial

Nature, Culture & Technology

Vol. 5

Natural Chance, Artificial Chance

Edited by
Massimo Negrotti

PETER LANG
Bern Berlin Bruxelles Frankfurt am Main New York Oxford Wien

Bibliographical information published by Die Deutsche Bibliothek
Die Deutsche Bibliothek lists this publication in the Deutsche Nationalbibliografie; detailed bibliographical information is available on the Internet at ‹http://dnb.ddb.de›.

British Library Cataloguing-in-Publication Data: A catalogue record for this book is available from *The British Library,* Great Britain.

Il presente volume è pubblicato nell'ambito della Ricerca 'La cultura dell'utente. Cittadinanza, mercato, tecnologia', finanziata dal M.I.U.R. e dall'Università degli Studi di Urbino 'Carlo Bo', PRIN 2004, Prot. n. 2004149922.

This volume is published as part of the research project 'The culture of the User. Citizenship, Market, Technology', co-financed by the Italian Ministry of University Teaching and Scientific Research (M.I.U.R.) and by the University of Urbino 'Carlo Bo', PRIN 2004, Prot. n. 2004149922.

Cover image: Massimo Negrotti
Cover design: Thomas Jaberg, Peter Lang AG

ISSN 1660-1084
ISBN 978-3-03911-476-4

Hochfeldstrasse 32, Postfach 746, CH-3000 Bern 9, Switzerland
info@peterlang.com, www.peterlang.com, www.peterlang.net

Printed in Germany

Scientific Committee

Yearbook of the Artificial

Published volumes

Vol. 1 Yearbook of the Artificial: Methodological Aspects and Cultural Issues, 272 pages, 2002.
Vol. 2 Yearbook of the Artificial: Models in Contemporary Sciences, 379 pages, 2004.
Vol. 3 Yearbook of the Artificial: Cultural Dimensions of the User, 249 pages, 2005.
Vol. 4 Yearbook of the Artificial: *Kyosei,* Culture and Sustainable Technology, 216 pages, 2006.
Vol. 5 Yearbook of the Artificial: Natural Chance, Artificial Chance, 284 pages, 2008.

Table of Contents

Massimo Negrotti

Chance would be a fine thing

Phrases! Phrases! As if it were not the comfort of all,
when facing a fact that cannot be explained,
when confronting an evil that consumes us,
to find a word that says nothing at all,
with which to soothe ourselves.

Luigi Pirandello, *Sei personaggi in cerca di autore*

...it is not astonishing that among an infinite quantity of matter,
which changes and moves incessantly, there should have appeared
the few animals, plants and minerals that we see;
just as it is not astonishing that in a hundred
dice-throws there should appear a pair.

Cyrano de Bergerac, *Voyage dans la lune*

Abstract:

Discussing the nature of chance is a different affair from that of a probability calculation. Defining chance means facing a fork in the road, as it were: on the one side, we have chance as the basis of probabilistic calculation; on the other, we must deal with a concept whose empirical content disappears just at the moment we wish to capture it. Since antiquity, the problem of chance has been investigated without reaching a definite conclusion: chance as ignorance, as unpredictability, as the lowest layer of reality, and so on. In many cases, as this paper seeks to show, the use of the word 'chance' could be avoided completely without losing the meaning of a statement. Nevertheless, in other cases, its use is required for invoking a sort of 'starting condition' of real processes that would otherwise remain without an initial scenario.

Although we are able even to reproduce randomness artificially, the ancient problem of its self-activation remains an open one.

Keywords:
chance, randomness, artificial, nature, variety, logics

1 Ambiguities and paradoxes

Not all of the words that we use designate shared concepts founded on experience. In this context, the word 'chance' is surely among those which, while being widely used, exhibit the least conceptual agreement among users, even if each of us would willingly testify to having encountered often, be it in professional or everyday experience, clear manifestations of the concept to which the word refers.

Already this first reflection leads us directly to the essence of the ambiguities, if not contradictions, that have always accompanied the term 'chance'; and furthermore, it points to a paradox that seems to me to be clearly evident. This paradox relates to the fact that, although we still lack a shared definition of *Chance*, and even find ourselves constantly in a state of deep perplexity when having to decide whether or not it exists, we nevertheless claim to be able to reproduce it artificially. Indeed, drawing of lots, sampling, simulations [Morgan, 1984], experimental strategies [Fisher, 1935], descriptions of statistical distributions, and a thousand other predictive and descriptive activities, have always made wide use of *Chance*, not merely as a convenient expression, but also as a property of the real world, and one that is measurable or even, when required, reproducible by means of various types of generator [Bennett, 1998].

Just what the presumption of reproducing an exemplar that we cannot define empirically might involve is not a subject to be discussed on this occasion [Negrotti, 2002]. It will be sufficient to add, however, that the paradox just indicated is certainly at the base of the age-old argument, between statisticians and mathematicians, scientists and computer scientists, on what should be intended when, for example, it is claimed that a sequence of symbols or numbers is, or is not, to be regarded as *truly* random. In fact, the statistical tests that we adopt, in order to establish the degree of randomness of a given sequence (and, in another sense, also the definition, suggested by Kolmogorov, of a random sequence as one that is not compressible into another, shorter sequence), are based on standard curves which are the effect of chance, and which do not, in the least, describe either the origin or the nature.

In any case, *Chance*, even if undefined, constitutes such a significant part of our conceptual arsenal, as it were, that, as a second paradox, it is often attributed with real causal powers. The expression 'by chance' signifies, in fact, the ascribing to *Chance* of a productive power, of a general nature, perhaps in expectation of the identification of underlying and as yet hidden causal agents. In essence, it is exactly this that is one of the notions of *Chance* that, since Laplace, has mainly prevailed: *Chance* as our ignorance of that which is occurring at the various levels of reality, and hence as a 'reservoir' from which originate, even if unbeknown to us, the true and ultimate causes of the phenomena.

On the other hand, considering the most advanced theories of modern-day physical and biological sciences, it would often seem that the expectation of more accurate tracing of the effective causes – thus shedding light in the darkness of randomness as our ignorance – is rather vain if it is true that, for example, in quantum mechanics or molecular biochemistry, the same unforeseen actions, variations or deviations are attributed to *Chance* as those that Lucrezio's *clinamen* indicated as a motor that generates, at an atomic level, the objects of reality, and therefore as an ultimate property of that which is real and not further analyzable.

2 Convergence of chance

The use that has, since ancient times, been made of the term *Chance* has undergone an internal differentiation with the birth of statistics and the mathematics of probability. Aristotle and Boezio, among others, tried to define the *Chance* we discover in single events (such as an unplanned and unexpected meeting with someone). As it was for Aristotle, an unexpected event is random, according to Boezio, when it is generated by causes that converge upon it even though they are preordained for other aims. This position, according to a rather exclusive perspective, is also common among modern-day scholars [Suppes, 1984]. On the other hand, identifying *Chance* with unpredictability [Eagle, 2005] may appear ambiguous. Indeed, while one can certainly assert that all random phenomena are unpredictable, it is by no means true that all unpredictable phenomena are random. At any level of observation, 'it must be understood that the purest determinism can give rise to the impossibility of prediction' [Jacquard, 1992].

With Laplace, rather, and then, above all, with Gauss, the foundations were set for seeing *Chance* at work in our descriptions, explanations and measurements of collective entities of phenomena by means of probability. Nowadays, the term *Chance* is thus adopted both in its Aristotelian version (which we shall call AC), for indicating an unexpected phenomenon, and in its quantitative version, probabilistic chance (PC), meaning that a series of events or symbols are distributed according to certain probability curves. The two conceptions of *Chance*, however, both tend to converge on the notion of unpredictability of individual events. The divergence of the two versions is therefore more apparent than real. Indeed, on the one hand, a single PC event belonging to a distribution (for example, the height of the next student to come into the classroom, knowing the average and variance of the given student population) is not predictable exactly, owing to its belonging to a population distributed according to probability. On the other hand, an individual AC event (for example, the sudden discovery of a chemical reaction thanks to the accidental collision of two flasks containing different substances) while presenting itself in an apparently isolated way, and therefore seeming amazing, nevertheless belongs also to a 'population' – namely, that which includes all events, even if heterogeneous and unmeasured, that are possible in a given context.

Therefore, *Chance* – conceived, provisionally, as a natural property capable of generating unpredictable variety – intervenes according to identical modalities in both AC and PC circumstances. The true difference between the two modalities, in essence, consists solely in the orientation of our observations. Observing a single event, *E*, produced by the unplanned encounter between two mutually independent events, *A* and *B*, we invoke *Chance* in an often psychologically relevant acceptation, solely because we avoid placing *A*, *B* and *E* in some wider context (the widest being the whole universe) in which those events occur frequently, or at least according to certain probability distributions. It is worth remembering that this last perspective was instead inaugurated by Chaos Theory, which underlines, precisely, the possibility of identifying causal chains whose initial conditions may be located 'a very long way' from the final effects being considered [Lorenz, 1993]. Accounts of everyday life, rather, tend to limit such causal chains to a few local events that are considered significant; a judgement, however, in the long chains of Chaos Theory, has just as strong a presence, even if upside down, as it were, because it tends to isolate a long line of events from the enormous quantity of collateral interactions with other unconsidered events within which such a line is inextricably contained.

Observing, instead, a single event, *E*, knowing that it belongs to a population of events of which we know the probabilities of occurrence, we invoke *Chance*, with or without psychologically relevant attributions, beginning with the awareness that its intervention was already assumed as a generator of a whole, known, distribution of possibilities that included *E*.

In other words, we know that if we choose a certain number on a die before it is rolled, the probability of a win is 1/6 (assuming that the die is not 'fixed' in some way). We can know this because we know the distribution of all the possible initial and final states of the system. By contrast, when, for example, while crossing the street for the umpteenth time, we find a precious object on the ground, we are inclined to isolate this event from everything else, as if, in essence, it did not belong to a context, and had no antecedents endowed with probability. It is due only to these psychological attitudes that, obviously, we make no mention of *Chance* when, upon crossing the street, we *do not* find anything of importance, or when we do not win at roulette. Yet *Chance* is directly involved in both cases.

3 Negative chance: a simple experiment

On the basis of the foregoing considerations, one might easily conclude that the word *Chance*, and perhaps the very concept we have so long sought to define, presenting itself as a common denominator in both its AC versions and PC versions, could be eliminated entirely from our vocabulary without any loss of information. We could limit ourselves, in other words, to considering that, with regard to the AC versions, a certain event simply happens, and with regard to the PC versions, the phenomena are always measurable only by dealing with the imprecision and uncertainty considered as properties both of our measurements and of reality. We could limit ourselves, in short, to suggesting that the events simply happen, according to probabilities that are sometimes known and sometimes not, without mobilizing any underlying 'generator' or 'cause'.

A simple experiment in this direction should provide for the elimination, from a set of meaningful propositions, of any expression containing references to *Chance*, in order to establish whether the set as a whole would suffer any deterioration. Table 1 can serve as a good example of such a simple experiment. It contains twelve, purely representative, propositions, taken from texts of a scientific nature, written for various occasions by persons of greater or lesser standing in the field of physics. Some of these propositions make use of the term *Chance* versions that are clearly of type AC, while others adopt the term in one of its PC versions. Its is noticeable that both these versions, while coming, in this case, from a specific field, are nevertheless in common use, not only in other scientific disciplines, but also in everyday life.

From the examination of the propositions, it seems plausible to conclude that the elimination of the expression 'by chance' would leave the information contained in them (and the associated meaning) by and large unaltered, with one particular difference which bears examination.

The explicit reference to *Chance*, be it in AC or PC versions, seems to carry out a single specific role: to describe an event as one that does not correspond to the plans or intentions of anyone or anything. A generic statement such as 'Yesterday I went to Milan, and, by chance, I met Mario at the station' would be rendered ambiguous if the expression 'by chance' were eliminated, as we would not know whether the speaker was referring to an unplanned meeting or a planned one. By contrast, a declaration such as 'Yesterday, while crossing the street, I found, by chance, a wallet' would not lose any information if the expression 'by chance' were to be omitted, because nobody crosses the street with the intention of finding wallets. It is reasonable to think that also when facing PC versions we would find ourselves, *mutatis mutandis*, in the same situation, because when drawing lots, for example, it would be completely superfluous to assert that a certain number had appeared by chance. In conclusion, recourse to the word *Chance* is made necessary when we wish to exclude, from our proposition, a reference to the known causes, including our intentionality, while such recourse seems utterly pleonastic when, from the content of the proposition itself, it appears obvious that the event could not have depended upon known causes or upon our intentionality.

However, even when the use of the term *Chance* proves necessary for the purpose off avoiding ambiguities, and thereby has a *negative* function, it appears obvious that its adoption suggests objectively that if the cause cannot be traced to intentionality, it must be found elsewhere, but its source does not appear accessible on the basis of available knowledge.

4 Positive chance: a logical necessity

If our knowledge of the empirical world were limited to a macroscopic phenomenology, definable as 'at a human scale', which would give rise to a highly anthropocentric view of things, then the matter of *Chance* could easily be dismissed as a pure question of convenience – as a mere expedient for encapsulating our ignorance in a word. But things have not been so simple since science has revealed microscopic physical and biological realities and, in essence, even if it is true – as it is true – that these realities, in pre-experimental terms, had already been suggested by the ancients.

What I intend to emphasize is that if we examine the most advanced knowledge currently available, *Chance* seems to be strengthening its presence, no longer only as a heuristic expedient, but also as a necessary, if unexplained, universal reality.

In quantum mechanics, from Heisenberg to Schrödinger, *Chance* has acquired characteristics that differ from those attributed to it in classical physics, wherein the recourse to *Chance* was, and is, always traceable to our ignorance of the phenomena at work beneath those that we intend to describe and explain. In quantum mechanics, by contrast, it seems inevitable to suppose that the particles themselves are able to make autonomous 'decisions' – as happens, for example, in Feynman's *random walk* of, curiously similar to the *wandering* of the atoms imagined by Lucrezio. According to this profile, the position of the School of Copenhagen, according to which *Chance* in quantum phenomena is intrinsic and fundamental, stamps a decidedly crucial full stop. According to a similar perspective, in general terms, '...nature seems to be made in such a way that the most important things in the real world appear to be complicated and accidental consequences of a multiplicity of natural laws' [Feynman, 1971].

Significantly, also in biology, in addition to the obvious reference to Darwin's theory of evolution, which places *Chance* at the centre of non-finalistic mutations (while anchoring the concept in question to our ignorance), the expression *random walking* is applied to the movement of bacteria, metazoan cells and amoebae [Banchetti and Ricci, 2002]. Although, in principle, one can speculate *random walks* at various levels of observation, and not only at a microscopic level, it is at microscopic levels that they pose the most serious problem: are they just a reflection of the limits of our cognitive abilities, and thus descriptive of the phenomena, or are they instead to be considered, at those levels, 'primary actions' of empirical reality? It is evident that if we could affirm that the microscopic levels at which science investigates are to be regarded as being at the *bottom* of reality, then we could deduce that *Chance* constitutes the final substratum of empirical phenomenology. However, with the current state of knowledge, this remains purely a hypothesis, even if, for various reasons, it seems somewhat arduous to proceed further without being able to deal with 'the wall of randomness', a sort of insuperable threshold in the face of which *Chance* would seem to be a property to be accepted as a fact.

5 Chance as an initial state

How one could imagine the 'intrinsic and fundamental' nature of *Chance* is certainly a problem of considerable, if not irresolvable, difficulty. As is known, Albert Einstein insisted that 'God does not play dice' ('I shall never believe that God plays dice with the world'), and it appears that Niels Bohr responded with the question 'Who are you to tell God what to do?' Steven Hawking added, very much later, that certain evidence suggested that 'God not only plays dice but also sometimes throws them where they cannot be seen'. Beyond the solemn judgements, it is a fact that in order to take a position regarding the 'existence' of *Chance*, we must always, at a certain point, take a position regarding its nature and its origin.

If we renounce this task, then we declare, implicitly or explicitly, that *Chance* is a mere construction of convenience or, in any case, a term that ends up applying only to our knowledge about probability distributions – knowledge that derives from the mathematical and statistical study of our measurements, and not from the verification of the phenomenon of *Chance* itself. The treatment that *Chance* receives from statistics, information theory, the mathematics of probability, and, at bottom, from the aforementioned AC versions of *Chance*, seems to assume exactly this position and then proceed upwards – that is, towards the formal study of the empirical distributions.

Conversely, the heated argument among the above-cited physicists – which, since the first half of the 20th century has extended to the world of more general scientific theory – regards, so to speak, *Chance* examined 'downwards'; that is, towards its definition as a supposed intrinsic property of natural reality. To sustain that 'God does not play dice' means, ultimately, to reject the idea that He has created a universe that is deterministically comprehensible only *locally* (in the macroscopic contexts in which the classical scientific laws allow valid explanations and predictions) but not in its entirety, or, if you prefer, in its full profundity. The problem, then, would appear to be the threshold below which microscopic phenomena begin and the classical knowledge of macroscopic phenomena therefore cease to be valid. Nevertheless, there would still be a need to explain how, even at macroscopic levels (in all empirical phenomena, from the physical to the economic, from the biological to the sociological), *Chance* makes its presence felt, in terms of describable events or sequences, only through the calculation of probabilities. The possibility should not be excluded, therefore, that the *Chance* which dominates microscopic levels is the same as that which generates effects at macroscopic levels, although, presumably, through some transformation, but without necessarily requiring the intervention of *hidden variables*.

What is certain is that when, in everyday research, we invoke some 'initial state' of a system, we do it with the conviction that we are able to provide an accurate description of such a state. On the other hand, it can be demonstrated that this is not necessarily true when the 'initial state' is established by an agent that exists outside the system made up of the observer and the system under examination [Svozil, 2000]. At a macroscopic level, to assume some statistical distribution as the initial state of a given phenomenon (for example, in an economic or sociological study, the distribution, in the various social strata, of the propensity to consume goods) entails accepting the idea that such a distribution describes a quantity, not further measurable, whose origins are lost, as already intuited by Laplace, in the maze of interactions with other distributions, known and unknown. In these circumstances, ultimately, it is assumed that the reality of the initial conditions is just as it is, and there is nothing more to be said.

At a microscopic level, by contrast, it is exactly the problem of the initial state that is crucial, and even if with regard to predictive qualities one can proceed by means of probabilities, the nature of the *Chance* which intervenes *right from the start* remains obscure – an 'initial situation of non-constriction' [Danchin, 1992] – as if it were, precisely, an intrinsic property of any innermost structure of matter. In other words, it seems that the nearer one approaches the root of things, the more *Chance* presents itself as something untouchable – that is to say, as something that one can only contemplate, upon which one can speculate but not isolate as a state of nature, because as soon as it is incorporated to some extent, it becomes something else, something determined and longer with any sign that indicates its provenance or genesis. All that would seem, then, to justify the magical or metaphysical notions with which man has defined, and now defines,

Chance: fortune, destiny, fate and luck [Noël, 1992] are indeed denominations that claim to close the discussion regarding the *cause* of the 'initial state' of an event or a combination of events.

In rational terms, the recourse to fortune or fate immediately appears in all its fragility, but, even entering into more analytic levels of discussion, *Chance* could be considered a sort of 'secular' version of fate or fortune, at least until we discover, if ever we do, which processes they generate in animate or non-animate things; that is, the sequences and events that the word *Chance* simply designates but does not describe.

Table 1

Text	Source	Version
'In February 1896, Henri Becquerel, conducting studies on fluorescence, noticed that uranium, even in the absence of phenomena that could be traced back to it, emitted new and unknown forms of radiation. The discovery, which happened **by chance** when the scientist found photographic plates exposed to uranium contained in a drawer, opened the doors to a vast and prolific series of studies.' (Original text in Italian.)	Marco Di Marco ***Archivio Nucleare*** *PROGETTO MANHATTAN* Chapter 1 – first part 21 November 2006 <http://www.archivionucleare.com/index.php/2006/11/21/progetto-manhattan-capitolo-1-prima/>	Aristotelian Chance (AC)

Text	Source	Version
'But a proton left to itself in empty space is endowed with this baryonic charge which it cannot get rid of unless it encounters **by chance** a wandering antiproton; it is endowed with a forced stability.' (Original text in Italian.)	Pio Passalacqua, *Breve Storia della Fisica delle Particelle*, <http://xoomer.alice.it/llpassal/ breve5.html>	Probabilistic Chance (PC)
'His interest in the biology of organ and tissue transplants sprang from his experience during his military service at Valley Forge General Hospital (VFGH) in Pennsylvania. As a lieutenant and with only nine months' experience as a surgeon behind him, he was assigned **by chance** to VFGH while waiting to be sent overseas.' (Original text in Italian.)	Anonymous article on Joseph E. Murray, Nobel Prize for Medicine in 1990. <http://www.hypothesis.it/nobel/ita/ bio/murray.htm>	AC
'As Pontecorvo observed **by chance**, the irradiated silver on those [wooden] tables became more radioactive than that irradiated on the usual marble table to be found in the same room.' (Original text in Italian.)	Ugo Amaldi, *La fisica dei nuclei dagli anni trenta ai giorni nostri*, <http://www2.sif.it/libri/fermi/ 10.pdf> Cited also at <http://www-news.uchicago.edu/fermi/Group25/vita_6.htm>	AC

Text	Source	Version
'Heisenberg had introduced this variable only as a formal artifice, and from a rigorously methodological point of view Majorana, at that time, was right in wanting to avoid the use of this new algorithm, but the subsequent history would show that, as sometime happens, Heisenberg's choice, perhaps **by chance**, or perhaps by brilliant intuition, was the right one.' (Original text in Italian.)	Edoardo Amaldi *Ettore Majorana, a cinquant'anni dalla sua scomparsa* <http://www.fisicamente.net/index-695.htm>	AC
'Enrico Fermi, **by chance**, did the same radiation experiment on different tables, of wood and of marble, and obtained different results, because the hydrogen in the wood slowed down some neutrons.' (Original text in Italian.)	Hans Bethe interviewed by P. Odifreddi, La Repubblica, 30 Settembre 2001	AC
'Synchronicities take the form of patterns that emerge **by chance** out of a general background of chance and contingency and hold a deep meaning for the person who experiences them.'	F. David Peat, *Synchronicity the bridge between matter and mind* <http://homepages.ihug.co.nz/~sai/ISSS_synchr.html>	PC

Text	Source	Version
'According to Poincaré [1913], the most important facts are those which can be used several times, or which have a chance to recurring. However, what are the facts of such properties? According to Poincaré [1913], the simple facts. Because it is evident that a complex fact is composed from many simple facts **by chance** and that only a still more probable chance could ever so unite them again, on the contrary to the case of a simple fact.'	Miroslav Pardy, *Creativity leading to discoveries in particle physics and relativity*, Masaryk University	PC
'On the university (nuclear engineering), things became boring quite often. I [thought] physics was over for me and that I would prefer computers, easy and instant reward. **By chance** during my military service, I read "gravitation" by Wheeler, then I read several books from the Landau series. These books were revelations for me.'	Michel, *PhysOrgForum Science, Physics and Technology Discussion Forums*, <http://forum.physorg.com/index.php?act=Print&client=printer&f=16&t=9754>	AC

Text	Source	Version
'Dirac's theory of particles and attempts to interpret this situation is Dirac's theory of particles and antiparticles, which may be designated, almost, as each other's mirror images. Both kinds of particles are conceived as arising through the formation of pairs and as reciprocally annihilating each other. The world in which we find ourselves belongs, **by chance**, to the one kind of particles, among which sporadically occurring antiparticles are very quickly destroyed.'	The Nobel Prize in Physics 1959, Presentation Speech, Presentation Speech by Professor E. Hulthén, Chairman of the Nobel Committee for Physics	PC
'There was no longer any doubt about what I wanted to do with my life – physics. But the kind of physics it would be was determined quite **by chance**.'	Roy McWeeny, <http://www.learndev.org/People/RoyMcWeeny/Index.html>	AC
'However, said Einstein, if the object is small enough – say, the size of a pollen grain or a dust mote – then the atomic-scale impacts won't quite average out. Sometimes, just **by chance**, one side of the object will feel a few more impacts than the others. [...] The result, said Einstein, would be random movements that never stop: the Brownian motion.'	NASA and The Hubble Heritage Team, Space Telescope Science Institute/Association of Universities for Research in Astronomy. End Photo Caption	PC

Bibliography

R. Banchetti, N. Ricci, 'Il fenomeno del displacing nel mondo vivente', in *Systema Naturae*, 4, 2002, pp. 15-59.

D. J. Bennett, *Randomness*, Harvard University Press, Cambridge MA 1998.

A. Danchin, 'Caso e biologia molecolare', in Ė. Noël, *Aggiornamenti sull'idea di caso*, Bollati-Boringhieri, Torino 1992, p. 98.

A. Eagle, 'Randomness Is Unpredictability', in *The British Journal for the Philosophy of Science,* 56, 4, 2005, pp. 749-790.

R. Feynman, *La legge fisica*, Bollati Boringhieri, Torino 1971, p. 179, seen in M. Cini (ed.), *Caso, necessità, libertà*, Cuen, Napoli 1998, p. 23.

R. A. Fisher, *The Design of Experiments*, Oliver and Boyd, London 1935.

A. Jacquard, 'Caso e genetica delle popolazioni', in Ė. Noël, *Aggiornamenti sull'idea di caso*, Bollati-Boringhieri, Torino 1992, p. 123.

E. N. Lorenz, *The Essence of Chaos*, University College London Press (UCL), London 1993.

B. J. T. Morgan, *Elements of Simulation*, Chapman and Hall, London 1984.

M. Negrotti, *Naturoids. On The Nature of the Artificial*, World Scientific Publishing Company, New Jersey 2002.

Ė. Noël, *Aggiornamenti sull'idea di caso*, Bollati-Boringhieri, Torino 1992.

P. Suppes, *Probabilistic Metaphysics*, Blackwell, Oxford 1984, p. 32.

K. Svozil, 'Science at the Crossroad Between Randomness and Determinism', CDMTCS-137, Wien 2000.

Massimo Negrotti, LCA – Laboratory for the Culture of the Artificial, Università di Urbino

Denis L. Baggi

The Use of Randomness in the Simulation of Creativity

Abstract:

This writing deals with the problem of simulating creativity in a computer program. While most programs are deterministic and produce a certain output for a given input in a reproducible way, non-deterministic and creative ones produce something from scratch, giving every time a different output, sometimes continuously. It is shown that this apparent creativity, meant to imitate human inspiration, can be obtained from a series of random numbers subject to prescribed choices, within a limited set of possibilities that give rise to an infinite number outcomes. The article describes some examples of computer music, starting from some of the 18-th century up to contemporary ones, including some realized by the author.

Keywords:
random numbers, deterministic vs. creative programs, Monte Carlo method, Markov chains, computer science applied to music and jazz, standard IEEE 1599

1 Introduction

A debate has raged probably for millennia, about source and nature of what is generally referred to as *creativity*, i.e. that human ability to produce something apparently new and from scratch, in the arts as well as in science, and in any activity. This article is not meant to add another contribution to this debate, but only to explore how creativity has been more or less successfully *simulated.*

The attempt to *mechanize human activities,* including *thought processes*, is historically as ancient as the activity itself – as witnessed by all mythologies, written records from ancient Egypt, Greece, Western Middle Age and Asiatic cultures of all times.

This has reached new heights since that ancient activity known as *programming* – already known and used by the Babylonians [Knuth, 1972] – has become a full discipline thanks to the invention of the *electronic digital computer* last century. The device has made it possible, for the first time in history, to *automatically translate abstract thought into concrete action* performed by a machine. And this has involved the *process of creativity* as well, and of its *automation*, which will be examined in this article on the basis of some past projects that have dealt with *music.*

2 Randomness and Programming

For the sake of this article and without ambitions of analytical depth, a distinction will be made between two basically different types of computer programs:

A. *Deterministic programs,* which always receive an input of some kind, and produce a result that is always the same
B. *Non-deterministic and 'creative' programs,* which receive no input, or little input as *guidelines*, and produce a new result, different at each run.

Class A is a straight derivation from the mathematical concept of a *function.* It is, indeed, not expected that, say, the cosine of 60 degrees varies every time the function is used! The class also includes a variety of applications that are valued for their *consistency*, such as *calculators, word processors, spreadsheet, Web browsers, music sequencers* and *score editors, communication programs, compilers* and *interpreters, dictionaries*, and most of the applications found on a typical desktop.

The programs of class B are mostly research projects meant to show that computers can be used to simulate some human activities. There is even a standard to assess the quality of the creation, the *Tur-*

ing Test[1] [Turing, 1950], the effectiveness of which can be debated, but which has the merit of offering a clear frame of reference. Every time such a system is run, a different answer is obtained. Examples are programs that compose *poetry, music, stories,* draw *images,* find new *mathematical formulas, concept associations,* and some *interactive games.* The category includes systems that may be given some kind of limited input, as guidelines in the same way a painter receives an order for a portrait, a composer for an opera or a requiem, a jazz musician a harmonic grid on which to *improvise.* While not generally found on the desktop of a typical user – except for games – it is this sort of programs that are the most interesting for this discussion, because of the lien between creativity and randomness.

3 Creative Programs and Randomness

For lack of any better understanding of the human creative process and of a physical-mathematical model of this activity, *creativity* has been simulated thanks to a *random number*[2] *generator.* Its output is subject to stringent conditions, such as *fitting a set of rules* – if not, the number is rejected and the generator called again – or used simply to select among pre-selected choices.

Notice that this is not an invention of computer science and not dissimilar to some methods used in mathematics. For instance, *Monte Carlo methods* have been used to model the behavior of various mathematical and physical systems. The most famous is the computation of a definite integral for which there is no symbolic expression. In this case, the simulation consists of *randomly shooting* at a picture of the curve in a known frame, so that the ratio of the shots under the

1 In a nutshell, if a user interacting with a machine at a terminal with any kind of question cannot tell whether at the other end there is a machine or a human, that program passes the test.

2 For all effects related to this discussion, no distinction will be made between *true randomness* and *pseudo-random series of numbers generated by an algorithm,* since, as simple practical experiments can show, given a set of rules stringent enough, the result of the program does not depend on the quality of such randomness.

curve with the total approximates the value of the integral. In general, *random* or *pseudo-random numbers*, subject to some kind of constraint, are used, and the method has been applied to architecture, design, computer generated films, video games, special effects in cinema, business, economics, high-energy particle physics, and other fields.

The constraints are often related to the algorithm of *Markov chains*, series that consist of sequences of states for which the conditional probability distribution of a state in the future can be deduced only from the current state. This is the case for the output of *formal grammars*, for instance those that generate *computer languages*, ever since ALGOL up to present C and ADA. The relation to this discussion is shown by computer algorithms which attempt to simulate *semantics* thanks to a dexterous use of *syntax*, the ultimate Holy Grail of computer science, to push syntactic simulation of semantics as far as possible.

4 Creative Programs in Music

Of all artistic disciplines, music is probably the one that, more than any other, has quickly appropriated all results of technology as soon as they were available: from the first pipes with holes and some kind of mouthpiece to 20th century electronics and computers, passing through medieval carillons, 18th century fine mechanics for the piano and 19th century precise craftsmanship for woodwinds. It is, therefore, no wonder that, independently of the artistic quality reached, attempts can be found throughout the history of music to simulate musical activities such as composing, improvising and solving musical problems.

The choice of music, therefore, is not *casual* (no pun intended!). Not only is *computer generated music*[3] the pet field of this author, but it can be argued there have been more attempts at simulating creativity in music than in any other art and that they have been more successful.

3 *Computer Generated Music* in the name of a *Technical Committee* in the *Computer Society* of the *Institute of Electrical and Electronic Engineers*, IEEE CS, founded by this writer in 1992.

4.1 Mozart, Haydn, and the Würfelspiel

Even though it is likely that examples of automatic composition are much older than the 18th century – Babylonian clay tablets with algorithms come to mind [Knuth, 1972] – the procedures for automatic composition by Mozart and Haydn are well known and documented.

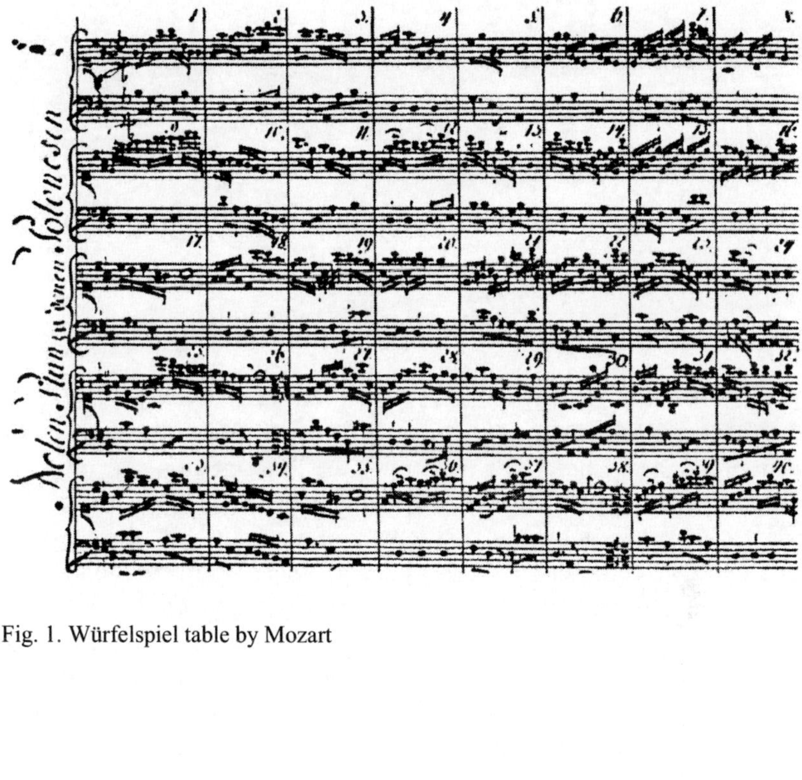

Fig. 1. Würfelspiel table by Mozart

Fig. 2. Würfelspiel table by Haydn

As the name says, the procedure is based on the outcome of *throwing a die* and it is therefore a variation of the Monte Carlo method with a random number generator. To simplify the description of the activity, assume a first page contains exactly 6 musical excerpts to choose from. Assume the 'composer' throws a die and hits a 4, he[4] then selects the 4th excerpt and moves to page 4. Then he throws it again, he hits say a 2 and thus selects the 2nd excerpt from the 6 on the page, and moves to page 2. And so on, until the piece is completed. Obviously the real procedure is more complex than that, because the more examples there are to choose from – e.g., with repeated dice throwing – the more varied the pieces. Mozart's table has 176 possible Minuet measures and 96 possible Trio measures to choose from, and with two six-sided dice used to determine each of the 16 Minuet measures (i.e. 11 possibilities for each of 16 measures), one die is used to determine each of the 16 Trio measures (i.e. 6 possibilities for each of 16 meas-

4 English being a sexist language, we are going to use the masculine form to simplify writing, without otherwise implying any hierarchy of sex.

ures), so in theory, there are (11^16) * (6^16) = (1.3 * (10^29) possible compositions [Mozart, 1787]. Likewise, in Haydn's work [Haydn, 1783], instead of what has been described above, there are separate tables with numbers, two for the minuet part and two for the trio, and those with music segments. The letters above each table indicate the 8 measures of the first part, A, and respectively of the second, B, while the numbers below each letter point to the music tables. Throw two dice, and then take the number under column A for the first measure: for instance, in the case of 6, look at row 6 and under column A you find number 48. Then, copy the music of this segment from the music table. If next time you get a 9, look at column B in row 9, which shows number 84, copy this segment from the music table. Keep going until you obtain 8 measures for the first part of the minuet, and in the same way obtain the 8 measures of the second part from the other table, and then the trio with one single die, from the tables for the first and second part.

Hence, the skill of the method resides in how carefully the different excerpts, tables and segments have been defined. Further, as in a Markov chain, a given sequence must concatenate in a meaningful way with the preceding and the next one, no matter how it was reached and what comes next. This is akin to the construction of a formal grammar to generate meaningful sentences in a natural language. The algorithm has been improved with techniques from artificial intelligence and symbol manipulation in program EMI, or *Experiment in Musical Intelligence*, which creates a new Mozart sonata from existing ones [Cope, 1991].

4.2 Illiac Suite

This project has the honor of being the first one realized with a digital computer, at the University of Illinois in Urbana Champaign in 1956, programmed in assembler on a Illiac computer [Hiller, 1959].

It consists of four movements, called *experiments,* for a string quartet, all of which are meant to investigate different compositional algorithms, to see what the music 'would sound like'. Thus, there was no pretense of reaching artistic heights, but with the clear aim to develop experiments possible only with the use of a computer. Only the second movement, the output of which is shown in Figure 3 and which deals of generation of four-part first-species counterpoint, will be shown here.

Fig. 3. Second 'experiment' from the Illiac Suite

There are eight sequences, which have been obtained as follows. Each random number, from 0 to 15, corresponds to a note in the key of C-major. Then, for each sequence from A to F, a new rule or constraint from *counterpoint* is introduced, as follows:

A. Random music; no rules
B. Skip-stepwise rule; no more than one repeated note
C. *Cantus firmus* starts on C with C chord for opening; cadence on C with leading tone in one of the four voices; resolution of tritone in VII-6, e.g. F/B must resolve to E/C
D. Octave-range rule
E. Only consonant chords permitted except for 6-4 chords; i.e., harmonic subroutine added
F. Parallel unisons, octaves, fifths and fourths still permitted; melodic subroutine added

G. Parallel fourths, 6-4 chords containing tenth still permitted
H. Best counterpoint

If the note does not fit the rules, it is rejected and the random number generator called again. This is therefore an application of the Monte Carlo method. Listening to the piece confirms the impression of randomness generally associated with computing machinery, notoriously incapable of grasping and conceiving a *whole* beyond locally correct details. Nevertheless, there is some merit in how, slowly, the music *converges* to something resembling 17th century music. The idea must have seemed so fascinating, over fifty years ago, that after century of dreams a machine has become available with unlimited possibilities for testing hypotheses.

The other movements are based on Markov chains and statistics, and are well described in the original book, unfortunately out of print, but available at many libraries.

4.3 Realization of the Unfigured Bass

This is, in its existing form, a fully deterministic program. It is possibly creative, but only in the sense that it creates something new from a given input by applying rules, and the result is always the same. The project consisted of automating the art of the *realization of the Unfigured Bass,* an exercise in tonal classical harmony known since the 18th century: on a given bass line, construct the *three upper voices*, *tenor, alto* and *soprano* [Baggi, 1974].

Figure 4 shows an example of bass the student must realize from a booklet of exercises [Pedron]. Unlike *figured* bass indications printed in 18th century scores for the harpsichord, no *figure* or *number* indicates the harmony the composer had in mind, and it is thus left to the student to find the most appropriate one.

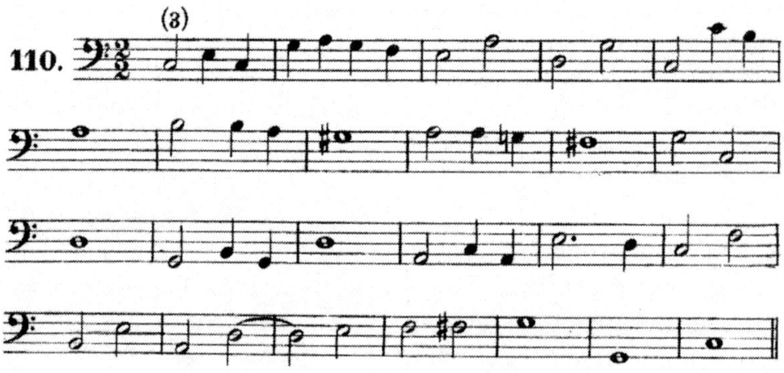

Fig. 4. Example of bass to realize

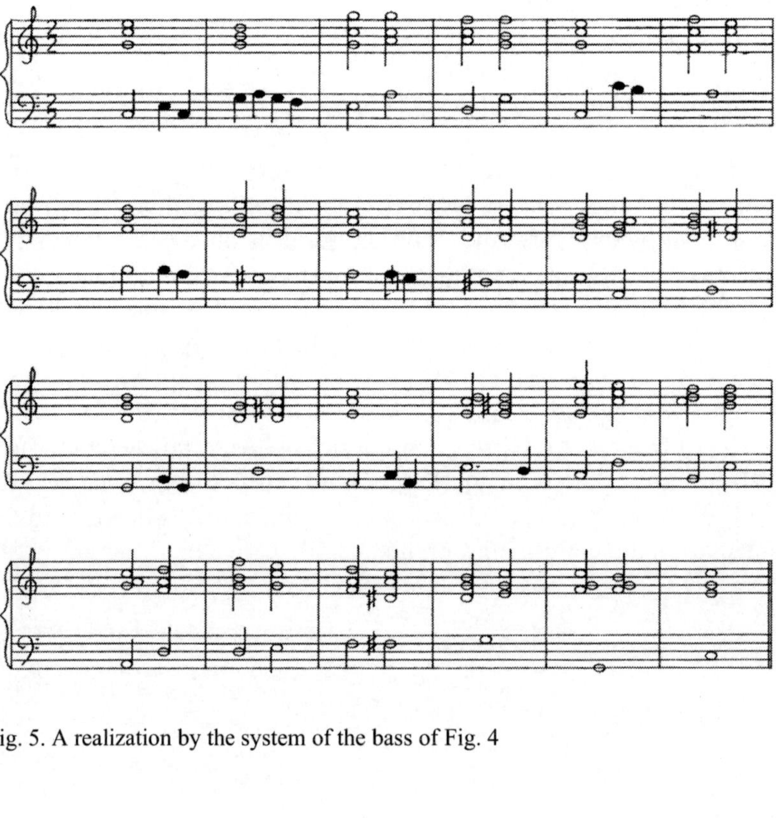

Fig. 5. A realization by the system of the bass of Fig. 4

Figure 5 is how the system *realizes* the bass of Figure 4, and for user's convenience it draws the resulting four-voice piece in a form mimicking that of a music score with a Calcomp plotter – no programs for score display existed in the punched-card, batch-mode programming of the time. The bass is clearly recognizable and the upper voices have been built 'by ear' by the system.

The program was fully deterministic because its aim was to demonstrate that the realization of the Unfigured Bass was possible at all, namely, that *at least one satisfactory solution would always be found.* However, in the same way that giving the exercise to a class of 30 students would produce 30 different answers, differing possibly in more than details, the system could have easily been built to strike different choices at critical decision points at random, thus producing a different answer at every run, and in particular:

- The system tried to minimize *modulations* and *tonicizations* to reduce the bookkeeping, doing so only when a note of the bass did not belong to the initial key; hence a sequence such as III – VI – II would not modulate (to the key of the relative minor of the subdominant)
- In the choice of harmony, the system stuck to rigid rules; hence, even though a circle of fifths in the bass was harmonized with 7th chords, there are many cases for which alternative harmonies would have been possible
- Except in progressions, the voice-leading algorithm attempted to keep all parts as close and common as possible from one beat to the next, and obviously many alternative choices would have been possible in many places
- Even though the model of four-part-harmony attempts to emphasize the harmonic quality of the whole, leaving aside considerations on esthetics, melody and the like, almost to the point of reducing the exercise to criteria such as right or wrong, attention could have been paid to the top melody, by adopting some of the rules of *counterpoint*, or embodying some criteria of melodic quality.

To sum it up, in spite of its limitations and of the 'primitive hardware' of the early 70's on which it ran – a CDC 6400 with 128k words of memory with a bulky LISP interpreter – the system was shown to be capable of passing the exams from the booklet, at some variable level of proficiency. Tonicizations and modulations are recognized, though only in the neighboring keys and relative minor-major modes. It is an example of *algorithmization* of four-part harmony, at a level at which criteria such as correct or incorrect do play a more important role than esthetic ones. And to conclude, it would have been easy to make this program truly *creative* by introducing randomness and the ability to create different versions of a piece, without breaking harmonic correctness.

4.4 NeurSwing

Pieces of jazz are rarely expressed with a *score* as in classical music. Jazz is based on *structures*, such as the 12-bar *blues* and the 32-bar AABA *Rhythm Changes* or *Anatole* [Baggi, 2001], which are supposed to be known and internalized, and *standards* which again are known both in harmony and melody. A new piece is expressed often only through its *harmonic grid*, a set of squares each of which contains a symbol expressing the nature of the chord it represents. Hence, the task of a *jazz rhythm section*, which consists in general of *piano*, *bass* and *drums* – with added guitar in the past – consists of translating this grid into actual notes played in real time with *swing* to support the soloist, singer or band.

Figure 6 represents an example of a net built by program NeurSwing from a harmonic grid. At the bottom, examples of alternative harmonies molded as a *neural net* are shown, representing alternate choices the whole system may choose from. At the top, a more conventional neural net is meant to illustrate the action of three *style controls: hot/cool,* to push the rhythm section in an expressionistic *hot* way, or *cool,* as it was popular in the early 1950's; *dissonant/consonant,* concepts borrowed from classical music, influencing the choice of more or less dissonant harmonies and chord parts; *as-isness/free,* which forces the system to adhere to the given input or to stray away in a more free way, still within the bounds of *harmonic equivalence*. There are, therefore, several choices at any given point, at the harmonic, rhythmic and melodic level by piano, bass and drums, which generate an altogether different output each time the system goes through playing one *chorus*. Hence, the same chorus is extremely unlikely ever to be replayed, and the system can go on forever simulating a rhythm section for a given user.

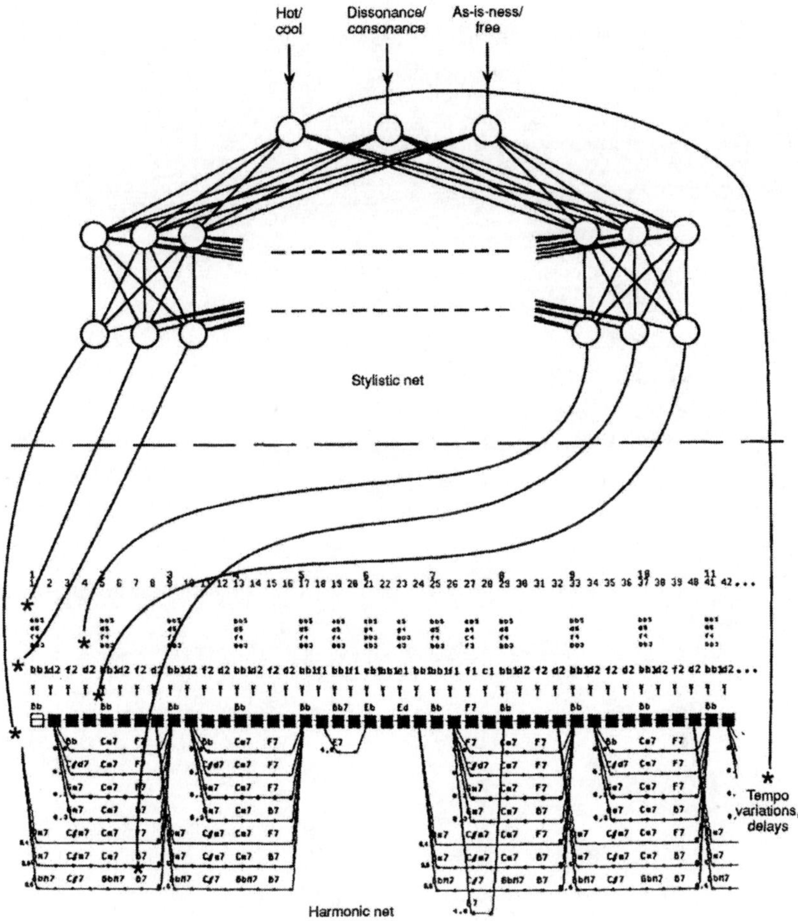

Fig. 6. Example of net built by NeurSwing

An example will illustrate the processing that goes on in real time. Table 1 represents the grid of a structurally elementary jazz sequence, 8 bars with the same chord, C-major:

C	C	C	C	C	C	C	C

Tab. 1. An elementary jazz sequence

which corresponds to the music indicated by a score as follows:

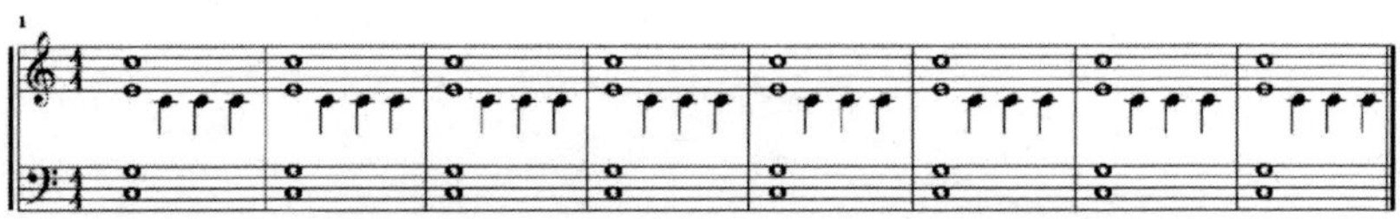

Fig. 7. Possible way of playing the sequence of Table 1

The music of Figure 7 can be seen to be extremely dull, even without hearing it, and is certainly not what an accomplished jazz rhythm section would play. Figure 8 is the score transcription of the output of NeurSwing from the input of Table 1, including the 2-bar snaps at the beginning and the use of a full drum set with high-hat, bass drum, snare, three toms, two ride cymbals and two cow bells:

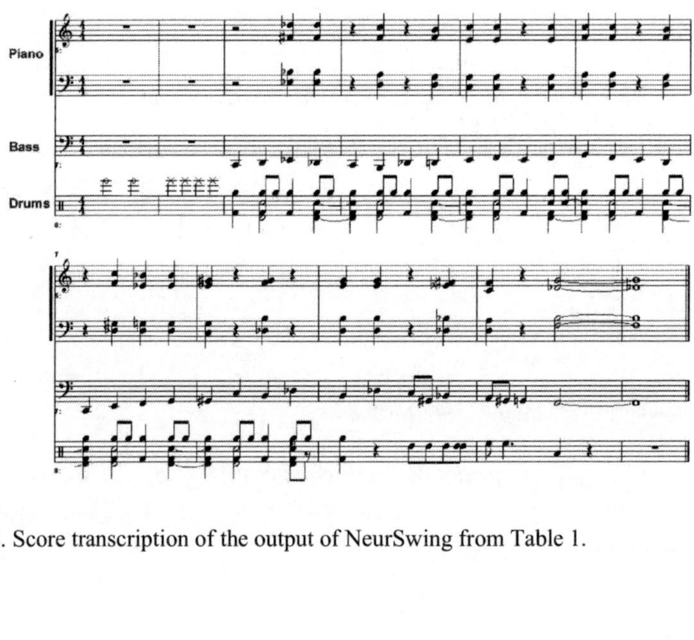

Fig. 8. Score transcription of the output of NeurSwing from Table 1.

Even though the impression of swing, which can not be annotated, would require this output to be heard rather than read, the variety in all instruments can be appreciated which contributes to making the input unrecognizable. Thus, the system acts as benchmark to experiment with various *models of swing*, at several levels: first, by acting on the *controls*; then, by *biasing the choices*; then, by introducing at the code level *alternative models for harmonic, rhythmic* and *melodic choices*; and finally, by *encoding different* and *non-linear choice functions*.

4.5 Standard P1599: Deterministic, but Creative Applications

For at least 40 centuries and in all cultures, music has been represented with *symbols*, which *archive the music*, store it as *notation*, indicate the way it ought to be *performed,* and gives indication on the *audio result.* The musical experience is much more than the passive listening to some binary file such as WAV or MP3. It is, instead, the act of *entering a new world*, *living an experience*, *understanding a narration* and *recognizing images*, with the possibility of *investigating how the whole is built* – which is the object of *musicology*, the science of music.

Standard *IEEE P1599*, under completion at SUPSI and at the Laboratory for Musical Informatics of the University of Milan, Italy [Baggi, 2005], defines how information is concentrated in a single XML file addressing *layers* [Haus, 2005], hence allowing reproduction of music in any format such as MP3, WAV, even MPEG, and MIDI, as shown in Figure 9.

All this can be illustrated with an example. Figure 10 is the screenshot of an application built at the University of Milan's Laboratory for Musical Informatics to illustrate the encoding of *King Porter Stomp,* composed by American musician and pianist *Jelly Roll Morton*, aka *Ferdinand Joseph La Mothe* (1889-1941).

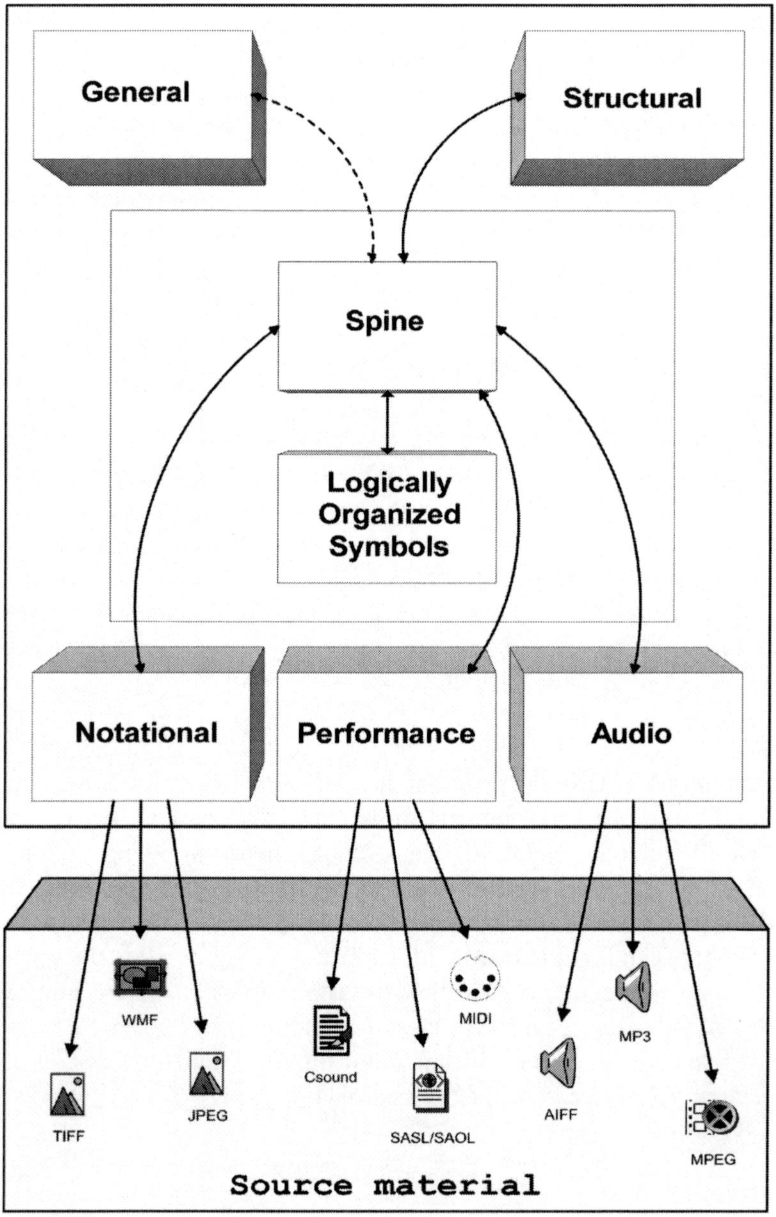

Fig. 9. The layers in IEEE P1599

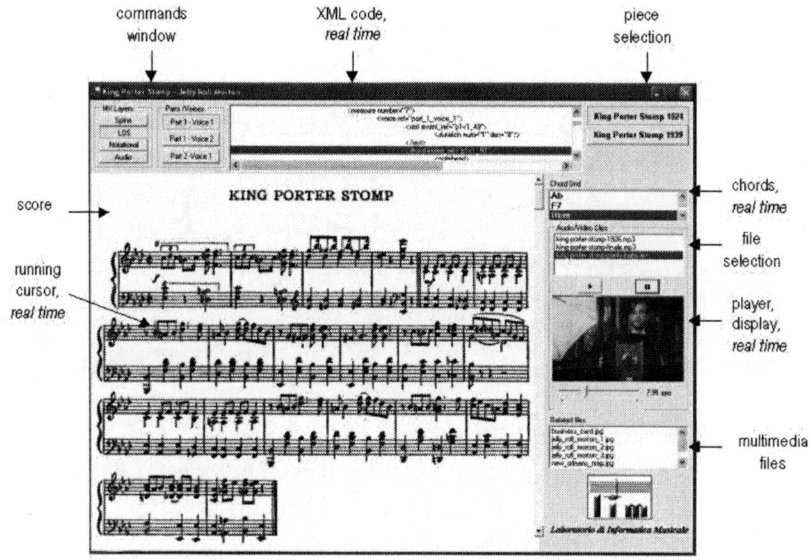

Fig. 10. Screenshot of an application for P1599

Beginning with the *piece selection window*, users have in this case two choices, King Porter Stomp 1924 and King Porter Stomp 1939, two published scores of that piece.

In the *file selection* window, the user can choose among various multimedia files. In this case, the choices include a 1926 recording and a MIDI version of the 1924 score, both encoded in MP3, and an excerpt from the 1978 film *Pretty Baby* by Louis Malle, a movie featuring a character, modeled after Morton, who is heard composing the piece in the background. The *player, display* window shows the movie, while for plain music it would instead look like a common non-video player.

When the user makes the latter selection, several synchronized activities start and execute, the movie segment starts and the music plays. All windows with the label *real time* in the figure operate synchronously. On the *score*, the *running cursor* indicates what is being played – in this case, the beginning of the sixth bar. The user can move the red cursor and restart play from another point in the score, while the other real-time windows and the player cursor adjust simultaneously.

The *XML code* window shows the encoded events of the *Logical Organized Symbols* scrolling with the music. In the *commands window*, the user can select which XML code is to be displayed: *Spine*, *LOS*, *Notational* or *Audio*, the elements of Figure 9, and select any of the three *voices* for the running cursor to follow.

The *chords window* displays the piece's harmonic elements, again synchronously with the playing, while the *multimedia files* window allows featuring of pictures that show portraits of Morton, his band, and other curiosities, such as the map of Storyville in New Orleans at the turn of last century, destroyed in the 1930's.

The above is a *totally deterministic* application meant to show the elements of the new standard. However, it is entirely possible to realize applications for which the display keeps changing and moving, and exhibits every time a different representation.

Figure 11 shows a proposal for a 'creative' browser that automatically displays varying images to convey the mood of the piece, in this case, the segment *Aum* in the suite containing also *Venus* and *Capricorn Rising*. While the music, represented by the central (blue) stripe, scrolls with musical symbols over an imaginary landscape, pictures appear that represent ancient Egypt – pride of the ancient past claimed by the composer – rising cosmological events, astrological charts, and the fallacy of a modern urban landscape, shaking while it hides social injustice. Hence, randomness has its place in this standard, and ensures that the user would approach the system every time with curiosity, to discover new aspects of the music, since the whole would look different whenever accessed.

In addition, there are several other types of applications made possible by this standard, with unending possibilities for queries at the historical, anecdotal, and musicological level, by embodying musical models and *ontologies* to derive concepts. Here are only a few examples:

1. *An opera.* A DVD of an opera allows the user to: see the play on the screen, hear the music, see the score (including manuscripts); read the libretto; choose excerpts of alternative renditions

2. *A piece by a jazz Big Band.* The harmonic grid is displayed and the name of the soloist pops up at the beginning of each solo
3. *A fugue.* The theme is highlighted as it gets passed among the different voices
4. *Music with a 'program' or story.* E.g., Vivaldi's Four Seasons come with poems that refer to segments of the music
5. *A piece of Indian classical music.* The scale of the raga is shown and the melodic development is highlighted
6. *A piece of several drums*, as in African Drumming, to show how the hits do not fall together
7. *Preservation of the music heritage from the past.* To store documents in any media [Haus, 1988]
8. *Musicological study.* Ease of queries – for example, all pieces utilizing the lowest note of a grand piano; questions as to why a certain note is used in a given harmonic context.

Fig. 11. A browser to explore record *Tauhid* by tenor saxophonist Pharoah Sanders, Impulse, 1969, LP and CD

5 Conclusions

The example above, in varying degrees, show how randomness, or random numbers, including pseudo-random ones, have been and can be used in programs that apparently *create* music from scratch, and come close to passing the Turing test, or at least illustrate some aspects of the art which is worth our while investigating. While human creativity is without doubts something much more complex that a simple filter of random numbers, since it is capable of focussing on a context and immediately exclude unwanted products, computer power is capable of offering an incredibly large number of possibilities, thus managing with brute force to compensate, at times, human ability.

Bibliography

D. L. Baggi, *Realization of the Unfigured Bass By Digital Computer,* Ph.D. Thesis, University of California, Berkeley 1974. Published by ProQuest Company, formerly Xerox University Microfilms, Ann Arbor, Michigan.

D. L. Baggi, (ed.) *Neurswing: An Intelligent Workbench for the Investigation of Swing in Jazz*, in *Readings in Computer Generated Music*, IEEE CS Press, ISBN 0-8186-2747-6 (case) – ISBN 0-8186-2746-8 (microfiche), 1992.

D. L. Baggi, *Capire il jazz: le strutture dello swing*, multimedia book with CD-ROM, Surveys of ICIMSI/SUPSI, Manno, Switerland 2001.

D. L. Baggi, 'An IEEE Standard For Symbolic Music', in *IEEE COMPUTER*, 38, 11, 2005, pp. 100-102.

D. Cope, 'Recombinant music: using the computer to explore musical style', in *IEEE COMPUTER*, 24, 7, 1991, pp. 22-28.

G. Haus, 'Rescuing La Scala's Audio Archives', in *IEEE COMPUTER*, 31, 3, pp. 88-89, IEEE CS Press, 1988.

G. Haus, and M. Longari, 'A Multi-Layered Time-Based Music Description Approach based on XML', in *Computer Music Journal*, MIT Press, Spring issue, 2005.

J. Haydn, *Musikalisches Würfelspiel,* original manuscript published in 1783. See also *Giuoco Filarmonico, ossia maniera facile per comporre minuetti*, G.Ricordi & C., Milan, Italy, N. 108202, year unknown.

L. A. Hiller, and L. M. Isaacson, *Experimental Music: Composition With an Electronic Computer*, McGraw-Hill, New York 1959.

D. E. Knuth, 'Ancient Babylonian algorithms', in Communications of the ACM, 15, 7, 1972, pp. 671-677.

W. A. Mozart, *Musikalisches Würfelspiel*, K 516f, 1787. See also *K.Anh. 294d*, Mainz, Schott's Söhne, Edition 4474.

C. Pedron, *Nuova serie di esercizi per lo studio progressivo del Basso senza numeri,* N. 14847, Carish S.A., Milano, year unknown.

A. M. Turing, 'Computing Machinery and Intelligence', in Mind, 59, 236, 1950, pp. 433-460.

Denis L. Baggi, SUPSI, University of Applied Sciences of Southern Switzerland

Danila Bertasio

And Narcissus returns to die in the arms of chance

The most beautiful of worlds is a heap of refuse thrown out by chance

Theophrastus, 3rd cent. BC

Abstract:

Art was once considered the science of sciences, able to reveal nature's secrets in the finest detail. But nature had other ideas, and was to prove a jealous guard of its greatest secrets. The clockwork precision that Bacon, Newton and others appeared to have revealed would, in the course of time, prove equally illusory – or, at least, intrinsically elusive. All of a sudden, insuperable limits were imposed upon observation and representation; and art – nature's self-appointed biographer – found itself with nothing to recount. Deluded and frustrated, art set off across uncharted waters in search of new treasures to bring home, but all too soon it was to find itself throwing overboard all that it had once valued, just to keep afloat.

Keywords:
art, anti-art, science, causality, chance, order, randomness, free will, Bacon, Newton, Happening & Fluxus, John Cage, Malevič, Duchamp, Maciunas, Pollock, representation, non-representational art

1 Introduction

All man's good and evil spring from his capacity to establish a balanced relationship between himself and the outside world. Observation allows us to collect information from our natural, social and cultural environments, and to construct representations, technology, theories, methods and actions able to guarantee our survival with a minimum of effort.

In the 18th century it was felt necessary to give some systematic order to this capacity; and thus, all that which found itself in the sphere of usefulness became science, while artistic activity fell into the sphere of uselessness. According to the mentality of the time, it was the laws of nature that determined events, and thus, for scientists, coming to know a phenomenon was no longer limited to providing a description of it, but also extended to discovering the laws that governed it. It was understood that only science allowed the identification of the causal connections underlying phenomena; and to the extent that one could determine the state of a certain system, whilst isolating it from external influences, it was possible to determine its future behaviour exactly.

In Germany, in those same years, university reform was put into effect. Science was no longer to be taught by philosophers, but by scientists instead. The institutionalization of science went hand in hand with a strong social legitimization, which, in turn, consolidated the alliance between science and the economy. The cultural conditions had changed, and even the second qualification of the Baconian system was realizable. All this would have implied an increase in production, and it was therefore worthwhile to support science enthusiastically, establishing it as a basic social institution. It was this context that led Comte to coin the expression 'scientific conception of the world'[1]. For him, scientific knowledge was intended as a form of authentic, 'positive' knowledge, and superior to any other.

1 This was a conception, however – as Agazzi writes – that was to be shaken to its foundations, between the late 19th and early 20th centuries, with the creation of non-Euclidean geometries, the discovery of contradictions in Set Theory, and of the inadequacy of classical mechanics with regard to its ability to absorb the theories of electromagnetic and thermodynamic phenomena, and to confront the subsequent revolutionary impositions of Relativity and Quantum Mechanics. It was then pushed further into crisis by an explicitly post-empirical conception of the world. (We need only think of the formulations accepted by the scientific community of philosophers of science, such as Kuhn, Lakatos, Feyrabend and Toulmin: cf. Agazzi, in Colombo, G. Lanzavecchia (ed.), *In Nome della Scienza*, Vol. 3, p. 395.)

It seemed that the times in which art was to be considered 'the science of sciences' had come to an end, and those who were still able to believe in art's function as a form of knowledge that was equally revealing of the laws of nature were reduced to but a handful. The awareness that for art it was impossible to arrive at an 'exact' representation of nature – that its representative results must remain anchored inevitably to a subjective representation of reality – manifested itself as a dysfunctional division between a so-called 'official' art, which survived thanks to a subdivision into various types, and an art which attempted to accentuate the transfigurative but subjective aspect of the artistic way.

Throughout the whole of Europe, the idea of the artist driven by an irrepressible impulse, in a state of mind dominated by fury and madness, was totally shared by the artists and writers of the *Sturm und Drang*, but opposed by the Academics. In short, art found itself battling on two major fronts: one *internal*, which exhibited an almost relentless break-up; and the other *external*, which undermined, from an exquisitely cognitive point of view, art's devotedly representative possibilities. Art's space within the cultural system was gradually being reduced, eventually finding itself constrained to the grassy borders of a byroad – almost like a dirt road in wide-open country: straight and narrow, and appearing to merge with the sky on the distant horizon.

The extreme attempt carried out by the artists who loved to paint *en plein air* then prompted their surrender to the idea of a reproducible 'outside world'. It was a point of no return, which culturally sanctioned an irremediable dualism between scientific and artistic forms of representation, and which led the artists themselves to abandon the ambition of intervening in the world. They thus began a sort of battle in exclusive defence of their creative and expressive freedom, yet freed from all claims of representativeness. This battle opened on two different fronts: one that upheld the necessity to operate a synthesis of all the arts, to go beyond its own expressive confines, to restore liturgical and sacred meaning to art, and which found an extraordinary prophet in Wagner; and another that pursued a sort of 'analytical' decomposition of every expressive form, according to which imagination and creativity could no longer be traced back to processes, and therefore to predictable results.

In essence, those who wished for a total art recognized their roots: they loved the mystery to be found in every artistic representation, and they did not disown their links with the art of the past. At about the same time, Debussy proposed a migration of the concept of *colour*, and the ideal encounter between painting and music performed by Kandinsky, in opposition to the anti-naturalism which seemed to him to live and breath in the culture of his time. Music, having abandoned for centuries the bending of its instruments to the imitation of natural phenomena, had conceded a greater space to interiority, and to a more careful search for new forms of sonority.

By contrast, far away from such conditions appeared, in fact, the choices of those who attempted the way of decomposition – those who tried an 'analytic' way within the scope of their own art. Picasso, in keeping with this perspective, took to stripping the figures and dismantling the faces of his women, almost as if their features were makeshift wrappings. That which he saw in observing the 'out of self' was nothing more nor less than a pretext for going further – or, rather, for asserting the plausibility of a radically alternative reality. Setting an *interpreted* reality against *observed* reality, the artist, a voluntary victim of a proud subjectivism, completed something that proved to be antithetical, with respect not only to the nature of the plastic arts up until that time, but also to the dominant culture of the time, which seemed able to bend the laws of nature to rational comprehension. It is interesting to observe, moreover, that while the science of that time was deluding itself, splitting hairs, attempting to arrive once and for all at closing, in the cage of 'good fortune', the inconvenience brought about by chance – by everything that defies prediction – art did not seem to be in the least injured by all this, and acquired faith in itself: an aware and free child, convinced that only 'looking inside itself' would enable it to find the right road once more.

The preconditions of Cubism were then amplified in Futurism, which analytically decomposed and dismembered, displaced and distorted the surfaces, letting objects melt into one another. Obviously, it was not the shareable appearance that drew the interest of the Futurists. The infinite layers of being appeared to become confused and fused in the objects depicted.

In the modern city, the difference between man and things has been reduced almost to the point of disappearance. Man in himself no longer arouses any interest; he must be abolished and replaced with matter. The heat of a piece of wood or iron is, for the Futurists, more fascinating than the smile or the tears of a woman.

The idea of taking a road that seemed to lead to a sort of 'universal life', made of unrecognizable levels to which the artist attributed the sense and significance that had been traced by the Cubists – and which, in the expectations of the Futurists, should have become broader still – appeared, in reality, to become ever more uncertain. In literature and poetry this was more evident still. The word became less transparent, widening the gap between itself and thought. Too occupied with denying the importance of the past to find space and sentiment for the new, it liberated itself from the eyes, the tears and the faces of women, and from trees and sunsets, remaining, instead, shackled to the electric light, to machines and motors, whilst forgetting that these, too, are a part of the world.

In 1915, Malevič, on the occasion of an exhibition entitled 'Second Futurist Exhibition of 0,10 Paintings', signed the Supremacist Manifesto, in which he proclaimed that the art of painting, of sculpture, and of the written word, were 'a type of camel loaded with all sorts of rubbish, with concubines, with Egyptian and Persian emperors, with Solomon and Salome, with princes and princesses in their gracious tutus, with hunts and luxuries worthy of Venus'. He went on to state that there had been no attempts, until that time, to undertake purely pictorial works, with no attributes from real life. Painting, for Malevič, was 'the bow-tie on the starched shirt of a gentleman, and the pink corset that squeezes the bloated belly of a fat lady'. It was the aesthetic side of the object, but it had never been an end in itself. Artists were 'bureaucrats, police officials' who conducted enquiries regarding 'perished goods, thefts, murders and tramps' [K. Malevič, 1916, p. 37].

Thus painting proclaimed its own indifferent autonomy from nature, ready to discover for itself, and in itself, new purposes – above all, with Malevič and his desire to abandon the paintbrush, a tool already considered incapable of penetrating the recesses of the mind; with Tarabukin and his conviction that representativeness was self-eliminating; with Mondrian and Rodčenko; and with the Stijl and

Bauhaus schools, which demonstrated a proud awareness of being reappropriated by their free will.

In this context, the French artist Marcel Duchamp made his seminal gesture. He took a wheel from a bicycle and remounted it on a kitchen chair, then signed the result. A work of art is an anonymous object which, removed from its designed context, is displayed by the artist in a museum or gallery. Implicitly, the object has become '*ready-made*', a common object, but chosen arbitrarily by an artist. Certainly Duchamp knew of the existence of the *collage* when he took to dismantling the bicycle; but what interested him was performing a 'transfiguration of the banal', and demonstrating that art is, before everything else, conceptual, a mental project. By displacing the object from its context and renaming it, one creates new thoughts that can be applied to it.

All this produces an effect of estrangement, also theorized in those same years by Sklovsky in the field of Russian Formalism: like Duchamp, the poet transposes all the symbols, becomes the instigator of the revolt against things, discards their old names and, just by renaming them, arbitrarily attributes new meanings, thus renouncing the poetic transfiguration of things in the name of the things themselves.

Not only was Duchamp not rejected, but he was legitimized in his way of doing art – exactly that which he, in justifying himself as a 'defrocked' artist, clamorously shunned, along with art itself. The object 'Bicycle Wheel' (1913) and the equally famous 'Bottle Rack' (1914) were immediately welcomed as artistic objects. (Unfortunately – for some, at least – both were later thrown away as rubbish by Duchamp's sister and sister-in-law when they were tidying up his studio.)

Since the advent of Duchamp, we have witnessed an enigmatic use of the term 'conceptual'. Yet we should not forget that every work of art is born of an idea, and is therefore, for all practical purposes, conceptual. This was certainly not an invention attributable to Duchamp, nor to those who would come after him. Is there, perhaps, a more conceptual art than religious art? An artist generates an object or a process that transfigures an idea that he cherishes in his own mind before externalizing it. The contemporary meaning of 'conceptual' is, therefore, only apparently revolutionary, because in reality it does not entail anything new.

2 What's new?

> Something must have happened at the beginning of the [20th] century; something that created the fortune of dissatisfaction. Everything that existed was no longer pleasing; nor was there any thought of improving things, or of recreating the sense of pleasure according to the classical models. Nature was no longer pleasing, nor were our homes; we no longer liked ourselves or our kinsmen. Beauty no longer pleased us; pleasure no longer pleased us [L. Fabro, p. 167].

Beginning with abstract painting, the age of informality had arrived: art included anti-art, and the artist, deceived into believing himself completely freed of the obsession of representation, assumed as his object the creative act itself, tacitly declaring a total indifference towards practical realization.

The illusion of making *being* and *action* coincide was dominant. Harold Rosemberg emphasises the point clearly:

> At a certain moment the American painters, one after the other, began to consider the canvas as an arena in which to act, rather than a space in which to reproduce, redraw, analyze or 'express' a present or imaginary object. The canvas, therefore, was no longer the support for a picture, but rather an event. The painter no longer approached the easel with an image in mind, but with the material in hand to transform, in some way, that other piece of material in front of him. The image on the painting would have been precisely the result of such an encounter [Rosemberg, 1964, pp. 14-15].

With Jan Foutrier, but above all with Jackson Pollock, a painting seemed able to consist in a random gesture with which to emphasize a total detachment from the long-established European tradition. However, their objectives were reached only superficially. Pollock, with his *dripping* technique, expressed, in essence, an empathetic conception of the act of painting, confirmed by his own theoretical declarations. Indeed, he himself claimed, for example, that he *controlled* the path of the dripping of colour when employing his *dripping* technique.

His was a procedure that was a throwback to the themes of the *einfuhlung* of European origin, thus confirming that the break with the past, while it may have seemed radical, was actually relative[2].

Once again in the U.S.A., along with the attempt to make creation and representation coincide, space was also found for the opposite approach: art brought forth a stereotyped and grotesque representation of *déjà vu*. The adherents of Pop Art and New Dadaism linked their art to the world of production and consumerism, feeding the market with icons of the most dramatic everyday banality. The content of *pop* culture has a deceptive fascination, because it is taken from the world of commerce, from newspapers, magazines, comic strips, publicity posters and television, and is adopted as the material upon which to perform aesthetic experiments.

Citationism, finally, through its gelid chromatism, reasserted the finite. Everything had been done: all that remained was for citation and *kitsch* to do, grotesquely, the reverse.

Art's aspirations have also changed: the artist can now enjoy success and notoriety not so much as a confirmation of his or her own expressive abilities, but through the ability to arouse astonishment, to transgress, to provoke.

2 Pollock, against his will, was unable to free himself from the representational dimension. The mathematician Henrik Jeldoft Jensen, of Imperial College, London, is one of several scholars who has attempted to explain Pollock's motives. His paintings are aesthetically pleasing because they obey fractal rules analogous to those of the natural world. As is well known, his expressive technique consisted in expanding lines of paint on the canvas in an apparently random manner, with a final result that nevertheless often presented an alarmingly naturalistic character. According to Jensen, Pollock's paintings have a fractal nature, and so they are not, in reality, truly abstract. It is known, for example, that Pollock would make the paint run backwards and forwards in waves by using a can punctured with holes, swung on a length of string like a pendulum. It has since been discovered that swinging a pendulum and hitting it with a hammer at just the right frequency (a little lower than its natural oscillation frequency) gives rise to 'chaotic' movement, leaving trails of paint on a canvas that can produce very convincing 'false Pollocks'.

3 'As beautiful as the fortuitous encounter of a sewing machine and an umbrella on an operating table' [Lautréamont]

On the basis of the foregoing developments, the step towards the improvisation of the untrained artist was shorter than we could ever have expected. *Anybody* had the ability to be an artist, should he or she but desire it. We can all be artists, as Maciunas arrogantly exclaimed in the 1960s, when exhibiting an enigmatic anti-art. Art is life, and the artist goes arm in arm with a modest common man who loves scandal and takes to randomly assembling objects, sounds and noises.

The attempt to eliminate, once and for all, the concept of *inspiration*, to deconstruct and decipher the artistic language, found the exaltation of the nonsensical in two exemplary *performances*: the first was the destruction of a violin, a very famous work by Nam June Paik; and the second was the dismantling of a piano, carried out by Philip Corner – all with the aim of highlighting the destruction of an idea of creativity that has always been associated with these instruments.

It was an entirely quarrelsome approach, but at the same time impervious to dialogue, and self-legitimized by assuming improvisation as the objective.

Maciunas, defining the programmes of the Fluxus, emphasized this clearly, admitting that:

> The forms of 'anti-art' refer, above all, to art as a profession, to the artificial separations between artist and public, between creator and spectator, between life and art: they are set in opposition to the artificial forms, to the models and methods of art itself, to the search for the purpose, form and sense of art: anti-art is life, nature, true reality – it is the one in all. The rain that falls is anti-art, the noise of a crowd is anti-art, a sneeze is anti-art, the flight of a butterfly and the movement of microbes are anti-art. These things are beautiful and deserve just as much consideration as art. If man could, in the same measure in which he lives art, experience the world, the real world that surrounds him (from mathematical concepts to physical matter), there would no longer be a need for art, or artists, or other 'unproductive' elements [G. Maciunas, 1962, p. 157].

But the critics and curators avoid this subject, perhaps because they desire to obtain valuable publicity in terms of journalistic notoriety and inflow of visitors to the shows. They show no sign of repentance, nor admission of error, perhaps because if they were to do so, they would be forced to question the sense of their own existence. In any case, there are no true critics: the so-called critics limit themselves to laudatory or celebratory comments, with no interest in allowing the public to notice that contemporary art has conserved nothing of the meaning of the terms 'artistic' and 'aesthetic'. This is nothing other than the expression of a confused subjectivism, no longer inspired by the 'masters'. And the artist, who now openly abuses the cultural significance of his title, remains still, quietly contemplating his navel, sustained in what he does by a system of art that survives by feeding its own presumption.

With regard to this point, we can glimpse the existence of a fine thread that links the artists of the early years of the 20th century to those of the more recent Second Avant-Garde: Picasso, Marinetti, Duchamps, Pollock and Worhol, arriving at Cage and the Fluxus. Together they have more or less deliberately given life to that hero of heroes, Oedipus, as he was created by Sophocles[3].

However, the analogy is soon broken: the artist, who, like Oedipus, has chosen self-blinding for his redemption, does not leave for a hypothetical Colono, but betraying his own desire for change, demanded so vociferously, continues to want to be called an artist.

3 Oedipus did not know how to resolve the paradox between free will and the chance that dominates it. He sought to understand the reason for the anger of the gods; to understand the evil that rained down upon him from the heavens. He wanted everything; he wanted too much. Reason was not enough for him, and so he consulted first the Oracle of Delphi and then the soothsayer Tiresia, but neither of them quenched his thirst for knowledge: he listened only to his own instinctive impulse. Oedipus wanted to gaze deeply into his image in the mirror, ignoring the fact that the soul is not reflected in the mirror, and that every reflected image does nothing more than recount the deception of the body, fatally eluding all else. Jocasta's suicide heightened his disorientation. Already aware of his own ambiguous guilt, in a condition in which innocence and objective responsibility showed fleeting confines, he left for Colono to atone for his involuntary sins, and there, in the sacred wood, he blinded himself.

The delicate thread of the analogy is broken just upon the difference between the extreme decision, which, in the case of Oedipus, is coherent, and the non-choice, sadly opaque and confused, of the artist. Oedipus went all the way, and chose to vanish into the sacred wood; the artist, for his part, has chosen to stop himself, preferring, as Theophrastus asserted in the 3rd century, to fool himself into thinking that the most beautiful of worlds could be 'a heap of refuse thrown out by chance'.

Different actions cannot but produce different reactions. And while sweet Antigone and the people wept sincerely for the loss of their king and father, the public looks to the art of the past, and not of the present, to satisfy its need for aesthetic transfigurations of the world. The blindness of the artist constrains Joe Public to become, on the one hand, ever more uncertain with regard to the utility and function of art, and, on the other hand, ever more mistrusting of a host of personalities, appearing more or less from nowhere, who, halfway between Crocianism and literary pretences, design for him, construct for him, define for him, just what he should regard as art. Artist and critic, profoundly united in the conviction that art is 'life' – and not the offspring of a pre-existing art – appear indifferent to the fact that the public is disorientated by the arrogant and anarchic idea of expressive liberty for one and all: a liberty that no longer knows how to offer any perceptual confirmation or discovery; a liberty that does not allow us to retrieve memories or imagine possible worlds, but concedes only a sort of complicity ambiguously improvised with chance – i.e., with all that which, as in life, acts unpredictably.

Unfortunately, we are not talking about an idea of *chance* that comes dressed in colours of good fortune, and which intervenes, almost like a spell, to resolve magically the representational problems of Protogenes. Pliny the Elder recounts:

> There is, in a painting, a wonderfully painted dog, like that which was, at one time, also the fruit of chance. Protogenes, indeed, recognized that he was un-

> able depict the drool at the panting dog's mouth, while he was satisfied with the rest of the painting, notwithstanding the great difficulties it had posed[4].

Rather, we are dealing with an idea of chance which, thoroughly troubled by figurative art, enters forcefully through the window, charged with all the ambiguity that even science has difficulty explaining. Indeed, we are dealing, in the words of Dubuffet, with 'all that which springs from the great black hole of little-known causes'. Dubuffet continues:

> [A]nd the artist finds himself grappling not just with any case, but with one case in particular, intrinsic to the nature of the material used. The word 'case' is inexact: it is better to speak, rather, of velleity and aspirations regarding a material that resists [J. Dubuffet, 1973, pp. 28-29].

It is also a material that renders exciting the leap into the unknown of the painter who begins to paint his own work, happy to die in order not to know where his adventure will lead him.; because, otherwise, once again in Dubuffet's words, 'his interest would fade if he already knew the result first and with precision; if he should create a work already conceived beforehand in his mind'. Thus the artist is 'coupled with chance'. Hans Arp, recalling his dissatisfaction with one of his unsuccessful designs, wrote:

> I continued to develop the technique of *papier collé* completely putting aside my will [...] and submitting myself to automatic implementation. I called this 'working according to the laws of chance' – laws that include all other laws; laws that escape us, just as the first cause that is at the beginning of everything

4 Pliny the Elder wrote: 'He had depicted the [dog's] drool according to the artistic rules, but it did not please him, precisely because it was only art; it could not be diminished and yet it seemed too much, and too far from the truth: one could see that it was painted drool, not exuded naturally from the dog's mouth. Worried and tormented in that moment, because he wanted a true likeness in the painting, and not just verisimilitude, he had washed away many successive attempts with a sponge, before changing his brush, without ever once being satisfied with himself. In the end, angry with art, saying that that it left him to understand too much, he threw the sponge at that cursed point on the painting, and it once again impressed the colours previously removed, producing exactly the result that he had desired all along. Thus, in that painting, chance created nature.'

> escapes us; laws that can be known only by abandoning oneself totally to the unconscious mind. I was convinced that anyone who followed such a law was creating life in its pure state [J. Arp, 1966, p. 328].

Arp's mistake, but also that of Dubuffet, does not consist in accepting, or not accepting, that chance is a characteristic that is intrinsic to nature itself, or in presupposing an idea of nature that maintains for itself a wide margin of liberty, but in the belief, rather, that through obscurity it is possible to express obscurity, that disorder is necessary in order to represent disorder, that the ineffable is preferable to the effable, noise better than sound, the mumble superior to the word, the formless finer than the form, to paraphrase Primo Levi. In other words, assuming chance as the *dominus* that nothing can oppose, one risks dying of boredom, or considering the artist's suicide as a legitimate alternative, as Dubuffet reminds us.

It should be emphasized, indeed, that there is a qualitative and sequential difference, which appears to have been overlooked, between free will and creativity. Suppose, for example, that an artist intends to represent his idea of beauty. He will exercise his own free will in choosing, from among many possible exemplars of beauty, the one that he considers the most suitable, and in conceptually identifying the characteristics that he intends to attribute to it. Indeed, we do not deny that an aesthetic fact, in its deepest epistemological sense, is accomplished beforehand in the mind of the artist. It is another thing, however, to attribute physicality to it – something for which the exercising of one's freedom of choice leaves the step towards the creative exercise, which is carried out in the attempt to unify the missing and physical elements of the expressive techniques, to modulate certain colours and forms, and to adopt composite constructive criteria: free will in the conceptual choice, creativity in its externalization.

But the modern-day artist holds that free will and creativity are simply synonyms; or, rather, that the former confirms the presence of the latter, which is flawed also from a strictly logical point of view. Rejecting the difference and the inevitable sequentiality between that which competes with free will and that which competes with creativity – a sequentiality that he can interrupt only with his surrender, with silence, with suicide, and so on – the artist foolishly thinks himself able to avoid incurring a transfiguration, and able, therefore, to produce representations. Unfortunately, however, he does not surrender and accept the risk of being transformed into a pathetic *bricoleur*, for whom:

> the main rule of the game consists in always adapting oneself to the means at one's disposal – that is, to a finite set of tools and materials, albeit heteroclitic, given that the composition of this set is not in direct relationship with the project of the moment, nor, for that matter, with any project in particular, but is the contingent result of all the occasions that have presented themselves to renew or enrich the *stock*, or to conserve it with the remains of previous constructions and destructions [Lévi-Strauss, 1964, p. 29].

According to the critics, ever more occupied and preoccupied with providing art with a seductive *maquillage*, suited also to assuring their own survival, John Cage introduced, into the musical and artistic context, the categories of probability, complementarity and indeterminacy: all categories produced in physics by the breaking of the deterministic cause-and-effect connection. The critics held that the reintroduction of chance, redefined, in its turn, as 'equivalence of possibility', would be the basis of knowledge for the Second Avant-Guarde. The Happening & Fluxus movement – of which Cage's work is often considered the origin, or, in any case, a sort of bridge that connects, at a conceptual level, Duchamp to Fluxus – is considered by many critics to be the area of thought in which quantal knowledge finds its most fertile soil. It is interesting to note that such analogies were 'ennobled' by the critics very much *ex post*, and not by Cage himself, who merely found it amusing to disorientate the public.

What has really happened is that the artist is no longer an artist. He has renounced exercising his ability in proposing new visual rules. He prefers to be raw intuition without rules, a grotesque hero who exhibits his own voluntary blindness, at times as a trophy and at times as a sacrifice to a confused ideology, propped up by a most faithful court, which helps him, day by day, as in the story of the king's new clothes, to wear an invisible garment.

The problem of contemporary art is perhaps to be found just here. In painting, unlike in writing, no attempt appears to have been made to keep under control that coefficient of uncertainty without which the certain becomes contrived and artificial. Mario Vargas Liosa, in *Lettere ad un aspirante romanziere* ('Letters to an Aspiring Novelist'), recounts how Flaubert used to find '*le mot just*'. The word was right *when it sounded right.* The right word was the one and only word that could fully express the idea. How would he know when he had found it? His ear would tell him. Flaubert would submit all his phrases to the proof *de la guelade.* He would go to read out loud what he had written, in a tree-lined street near Criosset's house, and his ear would tell him whether he had hit the mark or not. And it is just this that seems to be missing in the visual arts: the creative search for the '*forma giusta*'[5] that transforms literature, poetry and music into true art.

Tommaso Landolfi, in his works and in his diaries, is incessantly torn between the desire to be able 'to write randomly' and the acknowledgement that, as happens in reality, chance cannot be grasped by literature, because literature and chance are antinomical: the first combines, orders and systematizes, while the second positions itself, in essence, on the part of the unexpected and of disorder. In *La Bière,* Landolfi writes:

> Fatally my pen – that is, my pencil – turns towards a mastery of art: I mean towards a way of wording and composition that finally fought with the free writing proposed to me, and also with my desire to sidestep difficulty. Could I therefore never write randomly and without design, so as to at least glimpse, through the confusion and disorder, my deepest self? [T. Landolfi, 1991, p. 575.]

Landolfi gave up satisfying his desire, choosing literature and not silence, yet with an awareness that he could glimpse neither turmoil nor disorder, nor his inner depths. And it is exactly through such deliber-

5 To underline the importance of the search for the '*mot just*', Italo Calvino, in *Le lezioni americane*, wrote: 'As for the poet in verse, so for writer in prose, success subsists in the felicity of verbal expression, which in some cases may arrive like a sudden bolt of lightning, but which, as a rule, is born of a patient search for the "*mot just*", for the phrase in which every word is irreplaceable, for the most effective and meaningful juxtaposition of sounds and concepts. I am convinced that writing prose should not be any different from writing poetry: in both cases there is the search for a necessary, unique, mindful and memorable expression.'

ate renouncement that he knew how to keep alive the mystery of the artistic act. However, while he knew how to resist, he accepted the fact that part of his own thought remained unexpressed and inexpressible. In the visual arts we are faced with an idea of randomness that naively flings its arms open wide to the void, to the unknown, which signals and solicits silence.

We must finally ask ourselves, however, whether the disappearance of a whole series of values and cultural meanings, which nobody even feels the need to substitute, is the true core of contemporary art, rather than the surrender to figurativeness in the name of abstraction. How should we view the loss of values in art? By adopting, it would appear, chance as a dominant variable of artistic practice. The past is dead and the present only just remains. With the death of the past die traditions, power, the authority of the fathers and of the Church, and with them the detachment of the past's contents. In such a sense, life and art seem doomed siblings: if life is entangled, then art is entangled; if life declines, then art, too, declines. We, most civilized citizens of the third millennium, are disorientated; we no longer know how to distinguish the good from the bad. A certain corruption impedes us. Life is corrupted, and art, which is life's mirror, seems to have lost its way.

Bibliography

L. Aragon, 'John Heartfield et la beauté révolutionnaire', in *Les collages*, Hermann, Paris 1935.

J. Arp, *Jours effeuillés: poèmes, essais, souvenirs 1920-1965*, Gallimard, Paris 1966.

G. Braque, 'Pensieri sulla pittura', in E. Frey (ed.), *Il cubismo*, Mazzotta, Milan 1917.

P. Capanne, *Entretiens avec Marcel Duchamp*, Belfond, Paris 1967.

R. Delaunay, *Du cubisme à l'art abstrait* (edited by P. Francastel), École Pratique des Hautes Études, Paris 1957.

J. Dubuffet, *L'homme du commun à l'ouvrage*, Gallimard, Paris 1973.

L. Fabro, *Arte torna Arte. Lezioni e Conferenze 1981-1997*, Einaudi, Turin 1999.

E. H. Gombrich, *Arte e illusione. Studio sulla psicologia della rappresentazione pittorica*, Einaudi, Turin 1965.

W. Kandinskij, *Lo spirituale nell'arte*, SE, Milan 1989.

P. Klee, *Théorie de l'art moderne*, Denoël-Gonthier, Paris 1980.

W. de Kooning, *Écrits et propos*, École Nationale Supérieure des Beaux Arts, Paris 1992.

T. Landolfi, 'La bière', in I. Landolfi (ed.), *Opere*, *Vol. I (1937-59)*, pp. 575 & 650-51, Rizzoli, Milan 1991.

C. Lèvy-Strauss, *Il pensiero selvaggio*, Il Saggiatore, Milan 1964.

S. Lewitt, *L'art conceptuel: une perspective*, show catalogue, Musèe d'Art Moderne, Paris 1989.

J. M. Lotman, *La struttura del testo poetico*, Mursia, Milan 1970.

G. Maciunas, 'Neo-Dada en musique, théâtre, poésie et beaux-arts' (1962), in *L'esprit Fluxus*, show catalogue, Musée de Marseille, Marseille 1995.

K. Malevič, 'Du cubisme au suprématisme, le nouveau réalisme pictural' (1916), in *Écrits*, Il Suprematismo, Abscondita, Milan 1977.

K. Malevič, *Suprematismo. Il mondo della non-oggettività*, de Donato, Bari 1969.

A. Malraux, *Il museo dei musei,* Bompiani, Milan 1959.

F. T. Martinetti, *Manifesto del futurismo,* in 'Le Figaro', 20 February 1909.

H. Matisse, 'Notes d'un peintre' (1908), in *Écrits et propos sur l'art*, Hermann, Paris 1972.

P. Picasso, *Propos sur l'art*, Gallimard, Paris 1998.

Pliny the Elder, *Naturalis Historia*, XXXII-XXXV, circa AD 77.

R. Rey, *Contre l'art abstract,* Flammaion, Paris 1957.

H. Rosemberg, *La s-definizione dell'arte*, Feltrinelli, Milan 1975.

A. Strindberg, *Le hasard*, L'Echoppe, Paris 1990.

Danila Bertasio, Dipartimento di Scienze della Formazione e del Territorio, Università di Parma

Aldo Celeschi

Chance: from problem to product

The task of science is, as we know, to discover facts; but a random collection of facts does not constitute a science.

Irving Copi, Introduction To Logic

Abstract:

This essay is intended to define the investigative fields in which chance plays a significant role. While the nature and application of chance have often been discussed throughout history, a hard and fast definition of randomness suitable for all scientific sectors remains elusive to this day, notwithstanding the recent advancements in science and technology. At the same time, the usefulness of randomness in numerous methods and models is well known by all. In essence, sequences of random numbers and symbols have an economic value, just like that of any other product. Medicine, physical sciences, engineering, games, insurance and all sorts of sampling techniques make extensive use of products of this type. Their quality – in terms of adherence to the definition of randomness – must therefore be guaranteed.

Keywords:
sample, chance, randomness, contingency, cryptography, hardware, software, deterministic, Gaussian distribution, Georg Marmaglia, Laplace, probability, product, simulations, statistics

1 Introduction: chance as a concept and chance in use

Chance as an object of study has, over the centuries, had various interpretations, which are, however, fundamentally reducible to just two definitions.

The first considers chance as the autonomous origin of an event in which no objective causes intervene that are even potentially knowable, in direct contradiction to any deterministic conception of reality[1]. The phenomenon is observed in its entirety, and seems to emerge without any apparent motive or known physical law that could explain it[2].

The second definition holds that events are random if their generative processes remain unknown, but only because they are hidden in the extreme complexity of reality. Here, chance comes to be recognized *ex post*, through its effects, but does not elude, at least in principle, the known rules and laws.

Epicurus was the first to theorize the objective existence of chance in a world made up of atoms (i.e., indivisible particles). Very briefly stated, atoms typically have an ability to deviate unexpectedly and spontaneously from their vertical paths of free-fall[3]. In fact, a random event, which is just one of the many that are possible, appears to us to be independent of any law or determining cause, without thereby escaping the reality of our senses, and it thus poses itself as a totality, as a real fact. Laplace, instead, posed the basis for an interpretation of chance as a product of our ignorance and the impossibility of tracing the causal processes; yet he was the first to suggest the possibility of responding adequately, by means of calculating probabilities, to the consequences of causal unpredictability.

In all versions of chance that have been suggested, the occurrence of unexpected events in particular situations in which all decisions are possible, and thus aleatory, is attributed to chance, seen as something extraneous to man's wishes. Often masked by synonyms already well established in current-day vocabulary, effects and consequences that

1 With the term *determinism*, scientists and philosophers define the necessary link between phenomena thanks to the principle of causality. Turning on the scientific thought and revolution of Galileo Galilei, determinism tends towards the elimination of final causes.

2 This is the case, for example, when throwing a standard, six-sided die. While we know that the probability is exactly 1 in 6 that any given face will appear, we have no way of determining, before the throw, which face will in fact appear.

3 Lucretius later called this random process 'clinamen' (literally a bias or inclination).

do not always enter into the sphere of the perceptible are thus attributed to chance. In the attempt to clarify the normal use of this concept, we may distinguish at least three different situations in which the synonyms of the term *chance*, all of which are the legacy of both everyday human experience and scientific findings, are adopted, in the absence of any logical justification, in expressive forms that are at times tautological and at times allusive to mysterious realities. We can therefore identify chance as (a) a fortuitous event (i.e., a *contingent, coincidental* or *unforeseen* event); (b) a generator of events that are not dependent on man's wishes (luck, *doom*, *destiny*, *fate*); or (c) a particular and special situation (an *occurrence*, an *episode*, a *circumstance*, a *predicament*).

While common usage adopts these terms without any profound distinction, according to the given circumstances and linguistic style, it is interesting to observe how the sciences, and the professions based upon them, adopt the term *chance* in accordance with a more or less tacit reference to the definition provided by Laplace – namely, chance conceived as *unpredictability*. This happens when the initial conditions are not known, as in the case, for example, of a dispersion pattern produced by a large number of projectiles fired at a target: so many are the variables that come into play, that we cannot even imagine a calculation precise enough to approximate the reality. The same applies to the objective impossibility of performing a calculation to predict, molecule by molecule, the behaviour of a gas contained in a receptacle. This fundamental impossibility is a consequence of Heisenberg's Uncertainty Principle[4], according to which not all observable quantities can be measured simultaneously with arbitrary precision. In other words, the accurate measurement of a given observable quantity generates uncertainty in the measurement of the

4 Expressed by W. K. Heisenberg in 1927, the principle is a consequence of the principles that regulate quantum mechanics. The use of the word *uncertainty* continues to be the subject of some debate, as it does not seem to be a faithful translation of any of several German terms used to express the principle. According to the award-winning writer Bill Bryson, Heisenberg himself generally used the term *Unbestimmtheit*, and the playwright Michael Fayn has suggested that *indeterminacy* would be a better term, and *indeterminability* better still [Bryson 2004: p. 188 footnote].

others. Even here, when we know, for example, that it is impossible to measure the velocity of a particle, and therefore to predict its position at a certain moment, we are driven almost inevitably to the use of statistics – not so much to predict *when* a certain event will occur, but rather to know the probability that it will occur at all.

The relationship between probability and the number of samples (the frequency) thus becomes fundamental, and this procedure assumes great importance, because as the sample frequency increases, it tends towards the probability (which we do not know) of an event. We recall that one of the postulates underlying probability theory asserts that:

For every possible event there is a probability, P_i, that the event will occur. If n is the number of times the event could possibly occur, and V_i is the number of times that it actually does occur, then as n increases, the ratio V_i/n approaches the value of P_i.

2 Measurement of chance

Cryptography, simulations and the optimization of stochastic processes all depend upon the 'high quality' of random-number generators. These generate a numerical randomness, which is a procedure that at first sight may seem a simple one, but in reality is anything but simple, as randomness is in strict opposition to sequentiality and determinism. For this reason, according to John von Neumann, 'Anyone who considers arithmetical methods of producing random numbers is, of course, in a state of sin'.

Randomness is at the basis of various applications for which the primary requirement is always to create lists or distributions of reliable random numbers – that is to say, numbers of good quality. A typical example is sampling theory, in which every object or subject that makes up the sample, with all of its characteristics, must have the same possibility as any other of being chosen from the *universe* of all possibilities (Marradi, 1989).

If we generate a sequence of random numbers, corresponding one-to-one with the elements (objects, subjects) that make up a sample, and we then associate them with a complete, previously numbered list (universe), we obtain a group of objects or subjects, each of whose index number corresponds to one of the random numbers previously extracted. An example should clarify this concept further.

A landowner wishes to check the health of his 245 apple trees, numbered successively from 1 to 245. To do this he uses a simple probabilistic sampling method. He decides that 25 trees, chosen at random, should be a sufficiently representative sample, and he makes use of his home computer to produce the following table.

A1 ▾ f_x =INT(RANDOM()*245+1)

	A	B	C	D	E
1	85	200	150	22	176
2	148	15	175	151	218
3	88	35	214	111	14
4	101	4	210	201	214
5	91	153	227	28	62

The table is created using a standard Microsoft® Excel spreadsheet. The landowner knows that the software includes a pseudorandom number generating function, which can be called by typing in =RANDOM() in the appropriate text-box (labelled f_x). This function returns a pseudorandom number between 0 and 1 (but not including 1 itself). An example might be 0.7358351. In order to produce numbers suitable for his purpose, the landowner adjusts the formula accordingly, to produce whole numbers (integers) in the range 1 to 245:

f_x = INT(RANDOM()*245+1).

As we can see, the table contains 25 numbers in the appropriate range, each of which represents the number of a tree to be included in the landowner's sample. The landowner can now go to the orchard and begin checking the trees, starting with No. 4, the lowest number in the table, and continuing with Nos. 14, 15, 22, 28, and so on, up to No. 227, the highest number in the table. He knows that, according to the applied criteria, each of the trees in his orchard initially had the same probability of being chosen[5].

For consistency we could say that the term 'number' could be replaced with the term 'figure' of 'symbol', because, as often happens, random-number generators, be they based on hardware or software, take into consideration only two states, equivalent, by convention, to 0 and 1. One of the ways of dealing with randomness is by means of a probability distribution, meaning all of the information that determines the possibility that each element has of being extracted and becoming part of the distribution. When we represent, in a graph, the elements that make up a uniform probability distribution, we are presented with a distribution that is considered 'good', and the graph in fact turns out to be perfectly uniform.

With the following example, we shall see how a random variable appears in which the (discrete) values are included in a given interval. The test consists of extracting numbered balls from a container. If there are N balls in the container, and x is a given ball, then the probability that x will be extracted is $P(x) = 1/N$. Thus, if we have 8 balls in the container, the probability that any given ball, x, will be extracted is $P(x) = 1/8$, which in this case represents a uniform probability distribution, as each ball has the same chance as any other of being extracted. The same procedure is valid for discrete binary variables[6], which are represented using the symbols 0 and 1. If, as by convention, 0 represents the impossible event, and 1 the certain or necessary event, then in this type of distribution each value will have a 1 in

5 There was, in fact, a potential complication that might have arisen. There might have been some repetitions in the table, and thus certain trees would have been assigned for checking twice or even several times, meaning that the total number of trees selected would have been reduced. For the landowner's purpose, a few repetitions would make little practical difference: he might end up checking only 22 trees instead of 25. Nevertheless, if he were to insist on the need to check 25 trees, he could simply produce a new table that contained no repetitions. The probability that any similarly generated table will contain no repetitions is a little better than 0.7, or 70%.

6 The probability distribution for a discrete variable, x, is defined by the pairs x_i and $P(x_i)$. The function is also called the probability function. It must be positive and the sum must be equal to 1. Thus $P(x) \geq 0$ and $\Sigma_i[P(x_i)] = 1$.

2 chance of being selected. Because at least one of the elements (0,1) must be selected, the sum will always be equal to 1.

A commonly encountered and frequently used probability distribution is the so-called Gaussian or normal distribution, usually represented as a 'bell-shaped curve'[7]. In such a distribution, the values that have a greater probability of being selected are those near to the middle of the curve, while those farther and farther away have an ever smaller probability of being selected. The distribution may be described by a formula that constitutes a bridge between the mathematical sciences and the real world[8]. A full statistical description is beyond the scope of the current work, and we therefore limit ourselves to mentioning that the form of the 'bell' changes according to a function of the mean value (μ) and the mean square difference (σ). If the number of measurements could be extended to infinity ($-\infty$ and $+\infty$) we would obtain a distribution that belonged solely to the realm of mathematics, an one which could only ever be approximated in the real world or in the various physical sciences.

A normal distribution constitutes, in essence, a probabilistic model used in statistics to reveal the probability that any given phenomenon will emerge. Representing a normal distribution graphically means describing the probabilistic structure that a random variable can assume, whereby to every X there is associated a probability that X_i should manifest itself (G. Corbetta, 1999, p. 521).

7 Karl Friedrich Gauss described the 'normal distribution' in relation to the study of the movement of celestial bodies. Also, L. A. J. Quételet published an article in which he compared chest measurements of Scottish soldiers with the heights of French conscripts, demonstrating how such data are distributed according to the now familiar bell-shaped curve. Known also as the law of errors, because it describes the distribution of the random errors relating to the measurement of physical quantities, it is used to describe all sorts of phenomena, such as the probabilities encountered in games of chance, and the distribution of projectiles fired at a target.

8 $f(x) = \frac{1}{\sqrt{2\pi\sigma}} e^{-\frac{1}{2}\left(\frac{x-\mu}{\sigma}\right)^2}$ where $-\infty < x < \infty$ and $\sigma > 0$.

The curve produced is also known as the *density of continuous probabilities* or the *density of normal probabilities*, or, more usually, *Gaussian distribution*. Most continuous phenomena demonstrate a normal or Gaussian distribution.

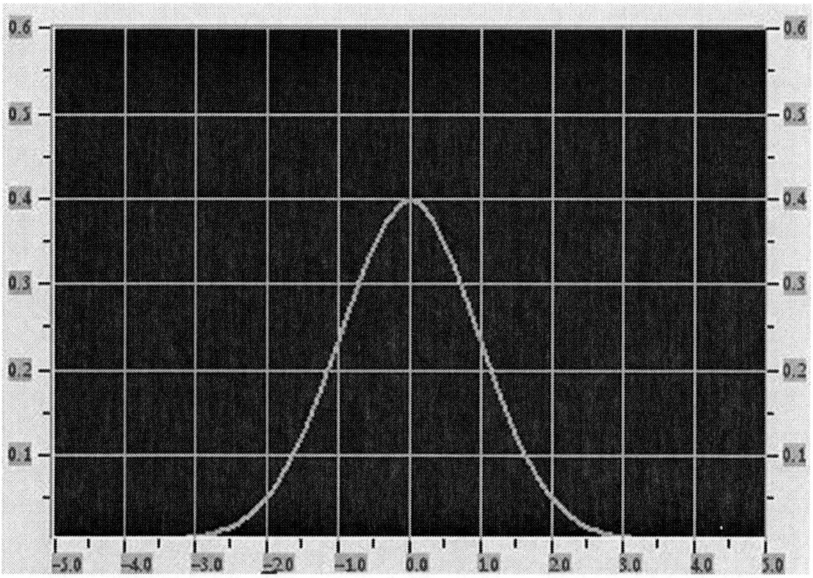

3 'Guaranteed' chance: producers and consumers

Just like any other commercial product, a distribution of random numbers, or the hardware or software that generates such a distribution, could be defined as an economic good, and as such it might offer a low, medium or high level of quality. Applications relating to specific scientific or productive sectors[9] need to make use of random distributions that are 'good or sufficiently reliable'. In the near future we may find, on the shelves of technical and scientific libraries, a series of

9 Here we intend software dedicated to specific areas of application, as distinguished from more generally applicable software. Some such areas may, in some way, not only necessitate dedicated software, but, for their correct functioning, require distributions of random or pseudorandom data whose 'goodness' will often determine the success, or lack thereof, of a given experiment or application. Such applications may range from those used in biology to those used to check, and perhaps predict, stock market trends.

suitably random distributions, verified by statistical tests, and 'guaranteed' by the creator of the algorithm or the designer of the machine that generated them.

We shall have, in short, as we already have in part, a type of guaranteed randomness, ready for use in fields as diverse as medicine, biology, agriculture, psychology, pharmacy, engineering and insurance. The professional sectors relating to such fields will be the natural 'customers' of this forthcoming opportunity (A. Bruschi, 1999, p. 553).

In this quasi-market, there are already numerous 'producers' competing to offer ever more reliable generators (Marsaglia and Aman, 1991; L'Ecuyer *et al.*, 1931; Lüscher, 1994)[10]. Among the hardware-based generators so far created, the best, or at least the most famous, remains that produced by the Rand Corporation in 1951, published as *A Million Random Digits with 100,000 Normal Deviates* (Brown, 1951). Various other attempts have since been made to produce generators that rely on all sorts of physical phenomena, such as electrical noise, background noise in transmissions (Brederlow *et al.*, 2006), cosmic rays or the radioactive decay of atoms.

The purchaser's faith in the 'high quality' of the distribution that he or she wishes to buy will be stimulated by the reputation of the producer and by methodological criteria that presuppose a standardization, which, however, has yet to be established. This last point is surely the most delicate of all, because it makes it necessary to establish certain parameters according to which chance can truly be guaranteed for successful insertion into the market. The primary problem is the type of distribution with which an algorithm works. Gaussian and uniform distributions in fact require criteria that differ according to the empirical situation to which they are to be applied. Consequently, the 'customers' will have to be aware, much more than they are at present, of exactly where their interests lie.

Offering a product of this type for sale means stimulating a sort of 'scientific' expectation on the part of purchasers, who will be inter-

10 George Marsaglia is also noted, in particular, for having proposed a battery of 12 tests, grouped under the name Diehard, used to evaluate the randomness of sequences of numerical values.

ested in evaluating the functionality and usability of the product, just as they would for any other instrumental good.

It should be emphasized, however, that in the circumstances described, the 'goodness' of the product is not objectively quantifiable if not by means of tests of the same statistical nature, adopted by the producer to evaluate the product. The validity of a given generator of random sequences, be it hardware-based or software-based, depends in principle, therefore, on the problem of providing a definition of 'randomness' that precedes the adoption of the tests, the choice of which, in turn, depends strongly on the particular school of thought regarding the intrinsic characteristics that a good random sequence should exhibit. This applies not only in scientific or professional fields, but also in everyday life.

The predictable value and the expectations of an investment are to be considered the driving force – the psychological spring, as it were – that pushes us, unwittingly, towards, for example, a *slot machine*, within whose workings chance appears to be guaranteed, and yet potentially reconfigurable on our part, thus rendering the machine our ally. While knowing that probabilistically speaking most spins of the reels will turn out in favour of the machine, the expectation born of the investment irrationally excludes all logical considerations, which are substituted by hope. In other words, complete ignorance of the functions governing the algorithms that generate randomness conspire with an acquired vice, to bring fortune to devices that were originally conceived to follow exclusively random processes. Thus we often find, among popular media sources, the sale of 'assuredly' winning numbers for the lottery, based on a randomness on the one hand guaranteed but, on the other, traceable with certainty according to a capacity possessed by few.

The nature of chance therefore remains a problem, notwithstanding the fact that it can be reproduced, for operational purposes, by means of algorithms or hardware devices. From a methodological point of view, the choice of a generator of random sequences of numbers or symbols remains, in turn, an operation that is in some way arbitrary and, in any case, not decidable *a priori*, because no statistical test can establish which among many sequences is the best, but only which of them is the worst. Thus, when chance is adopted for scientific purposes, everything depends upon the 'goodness' of the empirical results.

Bibliography

C. Arslan, *Statistica psicologica*, Edizioni Angelo Guerini e associati, Milan 1999.

E. Ballatori, *Statistica e metodologia della ricerca*, Margiacchi – Galeno Editore, Perugia 2004.

S. Borra and A. di Ciaccio, *Statistica – Metodologie per le scienze economiche e sociali*, McGraw-Hill, Milan 2004.

R. Brederlow, R. Prakash, C. Paulus and R. Thewes, 'A low-power true random number generator using random telegraph noise of single oxide-traps', in *International Solid-State Circuits Conference Digest of Technical Papers*, Jan. 2006, pp. 422-423.

G. W. Brown, 'History of RAND's random digits – Summary', in A. S. Householder, G. E. Forsythe and H. H. Germond (eds.), *Monte Carlo Method, National Bureau of Standards Applied Mathematics Series*, 12 (Washington, D.C.: U.S. Government Printing Office, 1951), pp. 31-32.

A. Bruschi, *Metodologia delle scienze sociali*, Edizioni Bruno Mondadori, Milan 1999.

B. Bryson, *A Short History of Almost Everything*, Black Swan, London 2004 (originally published by Doubleday, 2003).

A. Celeschi, *Informatica e metodologia della ricerca*, Edizioni Quattro Venti, Urbino 2001.

G. Cicchitelli, *Probabilità e statistica*, Maggioli Editore, Rimini 1984.

I. Copi, *Introduzione alla logica*, Il mulino, Bologna 1964.

P. Corbetta, *Metodologia della ricerca sociale*, Il Mulino, Bologna 1999.

L. Fabbris, *L'indagine campionaria. Metodi, disegni e tecniche di campionamento*, Carocci Editore, Rome 1996.

G. Ghiandoni and M. Rocchi, *Nozioni di inferenza statistica*, Edizioni Quattroventi, Urbino 1993.

P. L'Ecuyer, 'Random Number Generation', Chapter 3 of Elsevier Handbooks in *Operations Research and Management Science: Simulation* (edited by S. G. Henderson & B. L. Nelson), Elsevier Science, Amsterdam 2006, pp. 55-81.

P. L'Ecuyer, F. Blouin, and R. Coutre, 'A search for good multiple recursive random number generators', in *ACM Transactions on Modeling and Computer Simulation*, 3, 1993, pp. 87-98.

M. Lüscher, 'A portable high-quality random number generator for lattice field theory calculations', in *Computer Physics Communications*, 79, 1994, pp. 100-110.

A. Marradi, 'Casualità e rappresentatività di un campione nelle scienze sociali: contributo a una sociologia del linguaggio scientifico', in R. Mannheimer (ed.), *I sondaggi elettorali e le scienze politiche: Problemi metodologici*, Franco Angeli, Milano 1989, pp. 51-134.

G. Marsaglia and A. Zaman, 'A New Class of Random Number Generators', in *Annals of Applied Probability*, 3, 3, 1991, pp. 462-480.

P. Poletti and F. Vian, *Metodologia della ricerca e statistica*, Edizioni SUMMA, Padova 1990.

R. M. Ziff, 'Four-tap shift-register-sequence random-number generators', *Computers in Physics*, 12(4), Jul/Aug 1998, pp. 385-392.

Aldo Celeschi, LCA – Laboratory for the Culture of the Artificial, Università di Urbino

Mariella Combi

Chance as a cultural event

Each of us is a sort of crossroads where things happen.
The crossroads is absolutely passive: something happens there.
Other equally important things happen elsewhere.
There is no choice: it is a matter of pure chance.

Claude Lévi-Strauss

Abstract:

The theoretical background of this paper deals with 'chance' regarded as a cultural event giving sense to unforeseeable, social and environmental happenings. In particular, it concerns the characteristics and role played by chance in a global society ruled by uncertainty and risk. Among the important aspects of this analysis we can mention the relationships between local and global knowledge, between magic/religion and science, between people and institutions. In fact, everywhere and always, chance calls for an explanation that is apt to limit the risk perceived, and to reduce the incertitude it provokes. Processes leading to comprehension are required above all, but not exclusively, by unfavourable events such as unlucky incidents, accidents, failures, diseases, natural disasters, and so on.

Those who are struck by bad luck regard the event as fundamentally undeserved, which must be explained in some way. The notion of causality appears to be finalized to this end, in every culture, as well as to the social control through definition of what is reproachable. Whether magic/religious or scientific, the meanings given to causal events, firmly rooted in uncertainty, try to reassure the community passing through its members.

Keywords
chance, cultural event, cultural explanation, signification, local and global knowledge, uncertainty and risk, misfortune, magic, witchcraft, pre-literate communities

> In Zandeland sometimes an old granary collapsed. There is nothing remarkable in this. Every Zande knows that termites eat the supports in [the] course of time[,] and that even the hardest woods decay after years of service. Now[,] a granary is the summerhouse of a Zande homestead[,] and people sit beneath it in the heat of the day and chat or play the African hole-game[,] or work at some craft. Consequently it may happen that there are people sitting beneath the granary when it collapses[,] and [that] they are injured, for it is a heavy structure made of beams and clay[,] and may be stored with elusive [millet] as well. Now why should these particular people have been sitting under this particular granary at the particular moment when it collapsed? That it should collapse is easily intelligible, but why should it have collapsed at the particular moment when these particular people were sitting beneath it? Through years it might have collapsed, so why should it fall just when certain people sought its kindly shelter? We say that the granary collapsed because its supports were eaten away by the termites. That is the cause that explains the collapse of the granary. We also say that people were sitting under it at the time because it was in the heat of the day and they thought that it would be a comfortable place to talk and work. That is the cause of people being under the granary at the time it collapsed. To our minds the only relationship between these two independently caused facts is their coincidence in time and space. We have no explanation of why the two chains of causation intersected at a certain time and in a certain place, for there is no interdependence between them.[1]

The Zande wisdom also provides an explanation for the coincidence in time and space: it is witchcraft – otherwise the people would not have been there, or the granary would not have collapsed, as, indeed, had been the case up until that moment.

The example, cited by Edward E. Evans-Pritchard [1937, p. 69], proposes an anthropological perspective of the analysis of chance as a cultural event, an experience shared by human beings, who ask similar questions everywhere, and who always require an explanation.

With the term 'chance' here, I intend a fortuitous event, an unsought occurrence, an unplanned, unexpected fact that is posed, by human beings, as a central element of a process of explanation of forms of behaviour considered culturally significant for giving meaning to the world.

1 E. E. Evans-Pritchard, *Witchcraft, Oracles and Magic among the Azande*, Oxford University Press, London 1958, pp. 69-70. (Zandeland is a region of southern Sudan.)

Above all, but not exclusively, it is unfavourable incidents – misfortune, failure, disaster, illness, sterility, and so on – that require processes of signification: the explanatory choice made by the Zande society, with its witchcraft, provides its members with an explanation of the relationships between man and unfavourable events; with an immediate and predefined way of reacting; and with a system of values that regulate behaviour – 'What they explained by witchcraft were the particular conditions in a chain of causation which related an individual to natural happenings in such a way that he sustained injury'.[2]

If, for the Azande, witchcraft represents a dominant model of causality for misfortune, in more general terms it has long been maintained that some types of reaction to ill-fortune were a demonstration of the presence of a pre-modern mentality. According to these assumptions, modern man traces back the path from events to material causes, while pre-literate and traditional societies effect a path from misfortune to spiritual beings.[3] Robin Horton maintains that these societies tend to explain events by using causal inferences that have recourse to 'non-human agents', often ascribable to the magico-religious world. The notion of causality that acquires sense from these agents creates a connection between the world of real, everyday experience and the world of the extraordinary and unobservable. Moreover, traditional causal thought attributes, to a given effect, a simultaneous plurality of possible antecedents, according to the inclusive conjunction 'A and B and…'.[4]

A person who is struck by misfortunate retains it to be a fundamentally unjust event to which some sense must be given: the notion of causality, in all cultures, seems to have this purpose, in addition to social control through the definition of what is blameworthy or reprehensible. However, it also has aspects connected to forms of prediction, inasmuch as it aims at controlling uncertainty and risk. According to Mary Douglas, the definition of a given danger has the

2 *Ibidem*, p. 67.

3 Cf. M. Douglas, *Il rischio e la colpa*, Il Mulino, Bologna 1996, pp. 17-18.

4 Cited in N. Sindzingre, 'La necessità del senso: la spiegazione della sventura presso i Senufo', in M. Augé and C. Herzlich (eds.), *Il senso del male. Antropologia, storia e sociologia della malattia*, Il Saggiatore, Milano 1986, p. 98.

scope of protecting public welfare, while the attribution of blame means persuading other members of the community to follow the social norms.

For the fact itself of requiring an explanation – which, according to Nadine Sindzingre, implies a social referent[5] – an event is located within a process of causality in which it occupies a position as the *effect of something*. For example, an illness – which, from the point of view of causality, is indistinguishable from misfortune in some societies – is one of those events that unites mankind in the attempt to find an answer, a meaning.

Sindzingre conducted a study of the Fodonon people of Africa's Ivory Coast. The problem of causality among this particular Senufo tribe is, for example, difficult to dissociate from the empirical datum of physiological disturbance, which is, in turn, connected to an elaborately codified explanatory system. In the case of the Senufo, the identification of the relationship between a given symptom and a given cause is based essentially on a decision of a practical type. This allows the indispensable flexibility of the possibility of varying the function of the characteristics of the person involved, and of the social situation, including the interests and strategies of the community. That is to say, a particular symptom is not connected to a single cause, owing to the flexible character of the nosological categories adaptable to the real conditions present at the moment in which the event occurs.

In numerous African societies, the number of categories used to explain an ailment, and which are considered as possible causes, is practically infinite. In fact, the categories maintain certain preferential links that are represented, among the Fodonon, for example, by ideas regarding the body, both within and without witchcraft, by the metaphor of the falling of the spirits on their victim.

This plurality of causes is present among the Azande, and the choice falls, also in this case, upon the cause that is most suited to the social situation: it may happen that the cause can be identified more in common sense than in an external intervention, or in terms of natural causality rather than of mystic causality:

5 *Ibidem.*

> Zande belief in witchcraft in no way contradicts empirical knowledge of cause and effect. The world known to the senses is just as real to them as it is to us. We must not be deceived by their way of expressing causation[,] and imagine that because they say a man was killed by witchcraft they entirely neglect the secondary causes that, as we judge them, were the true cause of his death. They are foreshortening the chain of events, and in a particular social situation are selecting the cause that is socially relevant and neglecting the rest.[6]

In this way, they choose the cause that allows intervention and determines a certain social behaviour. Indeed, among the causes of an unfavourable event, incompetence, laziness, ignorance and stupidity of the victim of an accident or failure are all taken into consideration. The responsibility of the individual is present, as well as an awareness of the rational processes connected to the natural facts: first of all, they check that all the traditional rules have been observed – for example, by means of a technique that has been transmitted down through the generations – and only then will the cause be attributed to witchcraft.

The above-cited study of the Fodonon, among others, has revealed that the process of explanation is based on a fixed list of possible *a priori* causal categories to which they refer in order to find an explanation that simultaneously has both a predictive and a retrospective function. Often several such explanations are found. Possible *a posteriori* causes come into play in the interpretation of an ailment, based on divinatory practices. The polysemy of Fodonon causality implies that a causal category can act autonomously with regard to the content, while as a form it can be included in several of the categories considered possible. This characteristic structure shows the impossibility of formulating taxonomies. Douglas maintains, however, that we are not faced, here, with a lack of logic or intellectual capacity: the traditional theories are 'local' explanations that indicate pragmatic adaptations, as demonstrated by this population, which constantly recalls its own ancestry-based social organization. 'Causality founds its coherence[…] on the basis of the need for practical adaptations to real-

6 E. E. Evans-Pritchard, *op. cit.*, p. 73.

ity, which, being determined by *a priori* causality, are not destined to submit reality to discussion.'[7]

Every society has models with which to make sense of chance. Among the characteristics shared by the different models, there is a preoccupation with demonstrating the existence of links that are external to the subject who has been struck by misfortune. The externality of the causes must be 'demonstrated', be it by tracing them to gods, witches, taboo or the forefathers, or by appealing to biomedical or scientific knowledge. If no such causes can be identified, then, at least in our society, blame is apportioned, with the resulting social marginalization. This exogenous character of causality co-exists with other characteristics that enter into the sphere of 'common-sense' knowledge and experience, without excluding some natural causes that are the result of observation of regularities associated with phenomena.

Some sort of explanatory process comes into play in every society, generally unwittingly, particularly when dangers and risks are perceived towards life and limb of family members, lineage or society as a whole. The sense of uncertainty that these events arouse places causal connections at the centre of interest, and these connections create a strong emotive charge, precisely because of the fear that they provoke. In a socio-environmental context that is full of danger and uncertainty, anguish and insecurity, there is an increased recourse, as has been noted, to divinatory practices and the complex world of magic. Moreover, this happens even in western societies, even if this type of knowledge is considered false and irrational.

In pre-literate communities, for obvious reasons, there is a high probability that uncertain or risky events will occur, and it is in such societies that divinatory practices are most widespread. Yet divination should not be confused with the ability to predict future events. It is not only a technique of decodification, but also a cultural rationalization of causalities related to the order of the world and of mankind. The importance of the

7 N. Sindzingre, *op. cit.*, p. 113.

> intellectual and sociological phenomena that divination entails depends on the fact that they emphasize models of human action and types of more or less complex knowledge, whose configuration can both hide the maximum rationality of a society and emphasize, by contrast, the deviations.[8]

Recourse to the adoption of magic-based thought or scientific thought – or even of both of them simultaneously – allows the human species to 'control' the feeling of insecurity caused by various events: natural disasters, wars, internal power struggles, epidemics, famines, personal and collective misfortune, and so on. The sense of danger may grow or diminish, but is always present, to some extent, in the lives of individuals or in society as a whole. The prevalence of one or other type of thought is, obviously, in direct relation with the explicative and interpretative model present in the knowledge of the various societies, even if the attribution of randomness to the discoveries of magico-religious thought is certainly predominant. All forms of magic, notes Marcel Mauss, are enriched progressively by discoveries – some true, some false. The abandonment of a part of the magic that is *a priori* or irrational must be dealt with by the community. Magic

> approaches the sciences, which, ultimately, it resembles, in that it claims to derive from experimental research and logical deductions carried out by individuals. Magic resembles equally, and increasingly, the techniques, which respond, however, to the same positive and individual needs, and efforts are made to preserve only the traditional character of that which is collective.[9]

From another reflective viewpoint, the attribution of the inherent aim of giving meaning to random events divides, also, when one speaks of scientific knowledge and symbolic knowledge. The first is intended to determine nature's laws; the second, instead, to discover its meanings. In a thorough analysis of the role of this theoretical opposition, the anthropologist Roy A. Rappaport [2000] asserts that factual knowledge becomes fundamental when science divides from religion. Organized through classification, facts become theories aimed at generalizing, and thereby eliminating the existence of anomalous cases. Reflecting

8 J. Carlier, 'Divinazione', entry in *Enciclopedia*, Vol. 4, Einaudi, Torino 1978, p. 1226.

9 M. Mauss, *Teoria generale della magia*, Einaudi, Torino 1965, pp. 143-144.

on the perils of human existence, Rappaport emphasizes the importance of the gap between the natural law and the meanings that we construct: 'By the term "natural law" I mean to designate nature's regularities and their causes, whether or not they be recognized or understood.'[10] Moreover, it is difficult for the natural laws to be known completely. For the specific characteristic of human beings, who live through the meanings that they themselves construct, the natural laws are fundamental, but they cannot be separated from physical laws. The trick lies in the various cognitive methods that allow us to understand.

> If physical laws and the states of affairs they constitute are to be known, they must be discovered. In contrast, humankind's meanings must be constructed and accepted. Laws and facts and the scientific procedures for discovering them may provide some of the materials out of which meanings are made, but they do not, by themselves, constitute meaning, nor can they do meaning's work of organizing human action. The laws of physics, chemistry and biology, and the states of affairs contingent upon them, are the case whether or not they are known.[11]

The possibility of reconciling discovery and construction, continues Rappaport, represents an arduous undertaking, because a constant increase in our destructive capacities is accompanied by the crumbling of the certainty of our symbolic constructions.

Whether they be magico-religious or scientific, the meanings attributed to random events – i.e., events rooted in uncertainty – constitute an attempt to reassure the community, passing from person to person in order to reassure each in turn. Obviously, the overriding elements of signification for a society are dangers to life and limb, which are organized into explanatory models conforming to the model of the social structure. Mary Douglas recalls that in the 1950s, for example, nuclear energy aroused great interest and enthusiasm for technology: this explains the reason why a cognitive problem was perceived, between 'us' and 'them', relating to the 'knowledge of the true causes of things'. It was thought that science could enable us to rec-

10 R. A. Rappaport, 'Humanity's Evolution and Anthropology's Future', in R. Borofsky, *Assessing Cultural Anthropology*, McGraw-Hill, New York 1994, p. 154.

11 *Ibidem*, p. 163.

ognize the real dangers, which have causes that are objectively identified and theoretically demonstrated. Douglas writes:

> Chance, mystery and evil concealed themselves in remote corners not yet claimed by science, but, in general terms, thanks to our precise knowledge of the world, and to our powerful technology, our tendency to place the blame took us directly to the true causes, rather than deviating towards the supporting function of the institutions that developed elsewhere. This reasoning implied that, for us, something like a 'true attribution of guilt' would be possible.[12]

Interested in explaining the centrality of risk in the behaviour of modern-day Western society, Rappaport maintains that the adoption of a model of attribution of guilt is not carried out consciously. For example, some societies that differ from ours may favour the analysis of dangers in political terms, in the interests of the institution, while 'we' have separated the dangers from politics and ideology, placing them in a scientific perspective.

The current global situation may be described as being based on a new relationship between globalization, uncertainty and incompleteness. These three characteristics, which affect the whole world but are particularly explosive in our society, are instrumental in the analysis put forward by another anthropologist, Arjiun Appadurai, regarding the causes of ethnic violence in the last twenty years.

> Because of the growing multiplicity, contingency and apparent interchangeability of the identity available to individuals in the modern world, a sense of radical social insecurity is spreading, regarding persons, situations, events and norms, and even conceptions of the world.[13]

This uncertainty, which generates fear, anxiety and anger – to which Appadurai likens the incompleteness perceived from the point of view of identity – leads to violence towards 'cultural otherness', thus damaging the sense of belonging to a collective identity. The invasion of danger and uncertainty in everyday life no longer regards only strang-

12 M. Douglas, *op. cit.*, p. 22.

13 A. Appadurai, *Sicuri da morire. La violenza nell'epoca della globalizzazione*, Meltemi, Roma 2005.

ers or foreigners from distant lands, but also those with whom we share our daily lives, yet whose beliefs, habits and behaviour differ from our own. The suspicion arises that they hide, behind a mask, their true identity as traitors to the nation, intended as *ethnos*. The violence inherent in ethnic cleansing is founded upon this insecurity, which is socially sustained and politically instigated.

Another example relating to the world of chance is that of the game, intended here as the risks and hazards of life and experience, in the multiplicity of its disciplinary readings.

> The culture of hazard is grafted inevitably onto an existential condition strongly connoted by risk: to the measure to which insecurity and precariousness spread through life, and living itself becomes a risk, every choice, even the most rational and well-pondered, objectively touches the margin of hazard. It appears that our current epoch participates more than any other in this double nature of risk – as precariousness and challenge, as danger and adventure: risk has accompanied the greatest conquests that have constructed the civilization of which we are a part, and at the same time, in spite of the marvels of technological progress, it pervades more and more, in the forms of insecurity and fear, our daily lives, menacing our existence and fuelling our apocalyptic imagination. A sign of all this is the meticulousness with which present-day science continually constructs and updates the classification of risks, identifying, for example, the risks relating to food, the environment, health, relationships, economics, politics, and so on.[14]

Among the situations whose origins or significance are traceable to chance, another example is offered by some African populations, who use chance to explain the origin of the rite of circumcision as a search for beauty. The Ndembu of Zambia and their neighbours, the Bavole, speak of how chance gave rise to the rite of *mukanda*, the customary practice of circumcision. Some young boys, while playing in the midst of a certain type of grass that was sharp enough to cut, reports V.

14 D. Scafoglio, 'Il nostro azzardo', introduction to D. Scafoglio, *La vita in gioco. Antropologia, letteratura e filosofia dell'azzardo*, Merlin, Cava dei Tirreni (Salerno, Italy) 2006, p. 6. Among the papers presented in the same volume, many are concerned with interesting anthropological themes. Writers include L. Bonin, A. Buttitta, R. Brienza, M. Callari Galli and G. Harrison, O. Cavalcanti, and F. Romano.

Turner,[15] 'were accidentally circumcised', or, as another version of the story recounts, following the accidental cutting of the foreskin by the grass, the adults of the village completed the operation, removing the foreskin completely. Satisfied with the result, they tried to circumcise other boys and adult men: 'thus the people began to understand that it was better that all males be circumcised'.

Another version is that of the Dìì of Adamaoua (in the north of Cameroon), in which it is recited that

> one day, a woman saw some dog-headed monkeys cutting off their foreskins; the woman discovered that a penis cut in such a way was 'most gracious', and convinced her husband to let her circumcise him (Muller 1996:64). Her husband agreed and, together with the other men of the village, maintained that in such a way 'the penis had become more elegant'. The other men also circumcised themselves, but before proceeding, they killed the woman, in order that the 'secret' of the operation remain within the male circle, and that the women know nothing of it. Also here the fortuity of the discovery is rather evident, as is the presence of a woman who re-emerges at the origins of an operation that is instead purely male[...]. This is one of the 'numerous paradoxes' of Dìì circumcision; the other – we might add – is its aesthetic motivation, along with its randomness: so much suffering to make one's penis more gracious and elegant, according to a model discovered by a woman, by chance, and, above all, among some dog-headed monkeys? The fiction conceals the random origin, feminine and simian, of the circumcision; but its aesthetic motivation is completely clear.[16]

When reflecting on chance and its signification in the various cultural means of giving sense to the world, it is always important to keep in mind that meanings and knowledge not only reflect or refer to the real world, but participate in its very construction.

Bibliography

A. Appadurai, *Sicuri da morire. La violenza nell'epoca della globalizzazione*, Meltemi, Roma 2005.

15 V. Turner, 1976, cited in F. Remotti, *Prima lezione di antropologia*, Laterza, Roma-Bari 2001.

16 F. Remotti, *op. cit.*, pp. 109-110.

M. Augé, 'Magia', entry in *Enciclopedia*, Einaudi, Torino 1979.

R. Borofsky, 'Conoscere e conoscenza nelle attività culturali', in R. Borofsky (ed.), *L'antropologia culturale oggi*, Meltemi, Roma 2000. Orig. R. Borofsky (ed.), *Assessing Cultural Anthropology*, McGraw-Hill, London-N.Y. 1994.

J. Carlier, 'Divinazione', entry in *Enciclopedia*, Einaudi, Torino, Vol. 4, 1978, pp. 1226-38.

M. Douglas, *Risk and Blame*, Routledge, London and New York 1992. Italian translation: *Rischio e colpa*, Il Mulino, Bologna 1996.

M. Douglas and A. Wildavsky, *Risk and culture*, University of California Press, Berkley 1982.

E. E. Evans-Pritchard, *Witchcraft, Oracles, and Magic Among the Azande*, Oxford University Press, Oxford 1976.

C. Lévi-Strauss, *Mito e significato*, Il Saggiatore, Milano 1995 (1978).

C. Lévi-Strauss, *La pensée sauvage*, Plon, Paris 1962. Italian translation: *Il pensiero selvaggio*, Il Saggiatore, Milano 1964.

M. Mauss, *Teoria generale della magia*, Torino, Einaudi 1965 (1950).

M. Negrotti and F. Satofuka (eds.), *Yearbook of the Artificial*, Vol. 4, Peter Lang, Bern 2006.

R. Rappaport, 'L'evoluzione dell'umanità e il futuro dell'antropologia', in R. Borofsky, *Assessing Cultural Anthropology*, McGraw-Hill, London 1994. Italian translation: *L'antropologia culturale oggi*, Meltemi, Roma 2000, pp. 194-211.

F. Remotti, *Prima lezione di antropologia*, Laterza, Roma-Bari 2001.

N. Sindzingre, 'La necessità del senso: la spiegazione della sventura presso i Senufo', in M. Augé and C. Herzlich (eds.), *Il senso del male. Antropologia, storia e sociologia della malattia*, Il Saggiatore, Milano 1986 (1983).

D. Scafoglio, 'Il nostro azzardo', introduction to D. Scafoglio, *La vita in gioco. Antropologia, letteratura e filosofia dell'azzardo*, Merlin, Cava dei Tirreni (SA) 2006, pp. 5-19.

Maria Domenica Combi, Dipartimento di Studi Europei e Interculturali, Università 'La Sapienza', Roma

Mario Compiani

The ambivalent role of chance in modern scientific culture

Abstract:

The complementary roles of chance and determinism are examined in several branches of science. A concise diachronic analysis shows that chance plays a central role in modern science since it has been tamed by suitable theoretical tools. Nowadays it is a major ingredient of such diverse disciplines as the physics of complex systems, brain theory, immunology, genetics and Artificial Intelligence. Full acceptance of randomness in different realms is a particular feature of a larger structural change to a regime of limited rationality, culminating in the use of artificial neural networks in the description of very complex phenomena.

Keywords:
chance, determinism, genetics, chaotic systems, stochastic systems, evolution, genetic algorithms, artificial neural networks, memes, brain theory, immune system, structure of scientific theories

1 Introduction

In this paper we sketch the changeable role of chance as a critical category in scientific and philosophic culture.

We start from the ancient Greeks, who tried to expel fate and chance by means of divinatory practices, and we outline the trajectory up to the climax of the culture of enlightenment, in which chance was definitely pushed in the background. Noteworthy, a parallel tradition exists dating back at least to Lucretius, who assignes a constructive role to chance. But we have to wait until the end of the XIX century to see how randomness has definitely emerged from oblivion and how, in this framework, chance has got a positive connotation. The result is that nowadays chance has a two-faced role. The main aim of this paper is to shed light on the different faces of chance and to those cultural transitions which have led us to consider chance not only a dreadful menace to the ordered development of life, but also a useful cognitive tool that is a main ingredient of many modern scientific theories.

2 The decay of certainty

Tracing a short history of chance it is useful to start from the dychotomy casuality vs. determinism.

Looking backwards, it turns out that determinism is the corollary of a more ancient propensity to exerting absolute control. Determinism is the mere expression of a vital need, due to the fact that man strives to control the events of his life as he is threatened by the advent of the 'unknown', the unpredictable [Colli, 1980; Severino, 1980]. As Alexandre Koyrè put it, the valuable contribution of the Greeks to laying the foundations of western culture can be summarized as the transition from 'chaos' to 'cosmos' (i.e. from disorder to order).

In this connection it must be recalled that divination was much practised by the ancient Greeks.

Divination was the way in which the future and the obscure will of a god were made understandable to man by a designated mediator who had at all no control on the very process of communication and acquisition of knowledge [Colli, 1980]. The Pythia fell usually into an ecstatic condition to contact the god. Fate, in this respect, was the uncontrollable flux of events which man tried to contrast or anticipate.

Notably, predictability of the future situation entailed no constraint on the previous intervening states and, in particular, no limitation on the actual range of action, as strict causation of successive facts was totally unknown to the Greeks. Therefore to them there were no necessary causal links but mere intentionality, in the form of the arbitrary will of a god which had a conditioning effect on the future, albeit not at all definitive and irrevocable.

A different attitude towards chance is manifest in Lucretius. In his *De Rerum Natura* chance appears as the *clinamen*, that is to say small unpredictable deviations in the otherwise linear trajectories of the atoms. The *clinamen* causes the atoms to collide thus giving rise to the wealth of phenomena of the actual world [Lucrezio, 2001; Serres, 1980]. This amounts to ascribe a creative or constructive role to chance.

Both ways to look at chance were strongly redefined or supplanted by the rising parabola of determinism. The medieval christian vision of a universe ruled by the wisdom of God maintained that '*Nihil in mundo fit casu*' ('Nothing in the world occurs by chance'), as reported in an English picture of the XII century. The picture is reproduced and commented in [Saxl, 1985].

Just in the same way classical science has strived to expel casuality from its realm since the main requisite of scientific knowledge is reproducibility. Mention must be made of Musil [Musil, 1918] on modern science as the realm of repeatability '… rational knowledge goes under the banner of repetition of the fixed concept which does not take into consideration any deviation'.

The central role of repeatibility is extensively addressed also by Popper [Popper, 1970].

As a matter of fact, the impressive climax of modern science was rooted in deterministic laws. The very essence of this development is epitomized in the well known statement of Laplace in his *Essai Philosophique sur les Probabilités* [1814] (perfect predictability of the past and future of the universe on the basis of the complete knowledge of its present mechanical state) [Laplace, 1987].

Since then rationality has been identified with scientific rationality, and rational explanations inherited for long a deterministic connotation. This accounts for the otherwise unexplained delay with which statistical theories have been studied in the methodological studies concerning the essential requisites of scientific theories. The exhaustive review by Wesley Salmon, concerning the formal requisites of scientific explanations, shows it took nearly fifteen years for the nomological-deductive model for deterministic theory to be integrated with the first formal characterization of statistical theories [Salmon, 1992].

The triumph of determinism entailed that the deterministic paradigm has informed speculations in areas of science other than physics, like the social and biological sciences. For instance, du Bois Reymond, the founder of modern physiology and advocate of the experimental approach in medicine, boldly announced that physiology was expected to attain in its own theoretical descriptions the same exactitude as the astronomical and physical sciences [du Bois Reymond, 1973].

An example which is lively debated nowadays is genetic determinism, whose tenet is the so called central dogma of molecular biology (one gene – one protein), positing that the individual gene is the unique responsible for phenotypic characters (at molecular level).

However such a genetic determinism is loosing its original solidity and the very concept of gene as the unique determinant is a focus of major discussion, especially in the light of the recent developments in the study of heredity. Organisms are endowed with a very complex regulatory system made up of regulatory genes that make the organism very sensitive to environmental or endogenous stimuli [Fox Keller, 2001]. Epigenesis is gaining terrain as we recognise that much emerges during the development and is not written in the genetic code. Borrowing Bateson's terminology this is part of a vast process in which lineal thought is being supplanted by a recursive way of thinking [Bateson, 1989, 1997]. Interestingly, in the classical view inspired by the central dogma, the genome is much like the *homunculus,* a kind of preformed and microscopic copy of the adult man which, however, prior to the developmental stages, contains a miniaturized copy of all the features of the fully grown individual.

Firm faith in the existence of social deterministic laws was professed by the founder of modern sociology, Auguste Comte, who likened sociology to a kind of 'physique sociale' [Horkheimer, 1966]. The same cultural spirit of that time inspired Emile Zola's narrative cycle of the Rougon-Maquart. In the preface to *La Fortune des Rougon* [1871], Zola openly declares: '*L'héredité a ses lois, comme la pesanteur. Je tâcherai de trouver et de suivre [...] le fil qui conduit mathématiquement d'un homme à un autre homme*'. [Zola, 1966].

3 Coming to grips with uncertainty

Chance and randomness were admitted as main ingredients of scientific models only since the end of the XIX century. Accordingly, the statistical approach to macroscopic or mesoscopic systems was viewed as the best way possible to deal with the inherently noisy dynamics of these classical (non quantum) systems [Serra, 1986].

At this point randomness is definitely tamed and enters the set of computable entities, although this was perceived as a compromise with respect to the full deterministic description so far advocated by the enthusiasts of the triumphing classical physics.

Between the end of the nineteenth and the beginning of the twentieth century, physics was impressed a new trend, which testifies a decline of the control/determinism inspired attitude. This is due to the advent of quantum mechanics, stochastic physics (originated from the study of random motions), and the study of chaotic systems. Let us concentrate on chaotic and stochastic systems that are relatively less described in the epistemological literature although they provide the underpinning of the physics of complexity.

A typical chaotic system is the earth atmosphere. As a matter of fact, the basic chaotic dynamics were discovered during the early studies carried out on simplified models of the atmosphere. The peculiar property of chaotic systems is reflected in the well known fact that weather forecasting is reliable for a short time lapse (order of few days). The infinite memory of the past proper of a deterministic system is definitely lost due to the chaotic dynamics that amplifies the inevitable uncertainty affecting our knowledge of the present dynamical state. The final result is that, though the system itself is deterministic, it behaves much like a random system. For this class of systems internal deterministic mechanisms give rise to a pseudo-stochastic behaviour (deterministic chaos) [Serra, 1986] so that one is forced to use a probabilistic description.

Chaotic systems show that, in practice though not in principle, casuality is rooted in the very core of classical deterministic physics which so far provided the paradigm of predictability and repeatability. The pseudo statistical behaviour of the system introduces a weaker form of predictability, i.e. predictability that was once referred to individual particles is now referred to probability functions.

The prototype of stochastic dynamics is the motion of brownian particles which, in solution, undergo continual bombardment due to the solvent molecules. Under these conditions stochastic systems use chance (noise) as the very engine powering the whole dynamics. In the special framework of brownian dynamics noise has a double a contradictory role: as a matter of fact it provides the kinetic energy for the continual movement of the particles but it also blurs their deterministic motion which the experimentalist is essentially interested in. It is also worth noting that in the theory of brownian motion noise is the forcing term that concisely sums up the uncountable contributions due to the individual molecular collisions [Serra, 1986]. Mathematically, chance is a unique term that represents the overall effect of infinite deterministic agents in an 'economical' way, following Mach's and Duhem's terminology. This introduces the notion that randomicity may be a useful instrument to represent in an effective way the unmanageable complexity of an underlying deterministic mechanism.

Both in the realms of stochastic and chaotic dynamics randomness emerges as the consequence of our incomplete knowledge of the initial conditions or the interaction forces, whereas deterministic description of the system is always the fundamental model or the limiting form one tries to attain, at least in principle. Stochasticity is viewed as a degeneration of the pure deterministic description which nonetheless, in many cases, may be instrumental at simplifying quantitatively the description of the deterministic system at hand. This point of view changed with Reichenbach who maintained that statistical laws are the primary form and the deterministic ones are derived as special cases from them [Bachelard, 1978]. Bachelard shares the same opinion as he states that determinism derives from the neglect of supposedly irrelevant processes, i.e. those uncontrolled processes that are thought to produce negligible effects [Bachelard, 1978]. Determinism is therefore viewed as the result of a simplificatory process and this determines a revolution in our hierarchy of dynamical representations in that stochasticity becomes a primary notion and determinism is reduced to the rank of a derived category.

The notion that randomness may play a primary and dominant role is at the heart of current speculations according to which design, regularity and repeatibility and predictability emerge spontaneously in networks of randomly interconnected elements due to self-organization processes. That chance (and selection) is the very source of design is the basic hypothesis underlying the investigations carried out by Stuart Kauffman [Kauffman, 1993] which are in their turn inspired by the same spirit of the work of Jacques Monod's [Monod, 1972]. Similar mechanisms are readily applied to metabolic and genetic networks and might probably contribute to the current discussion on the so called Intelligent Design.

4 A new role for chance

Quite remarkably we owe some exponents of the 'other culture' (according to the classification of culture due to Charles P. Snow [Snow, 1963]) the intuition that randomness may be used as a cognitive tool and may have a constructive role. For instance, Henry Miller wrote that '… chaos is the score in which the real world is written …' [Miller, 1967] but Jonathan Swift is probably the author who better anticipated the notion that random search may be a useful and powerful strategy. His 'text-producing engine' was a device that could be used by the scholars of a remote country visited by Gulliver to produce random texts. It was conceived as an array of elements carrying the whole ensemble of the words of their own language. By operating the handles of the machine one could fix the output (a definite word) of each of them. This allowed one in principle to explore all the conceivable combinations of words so as to produce all the possible texts compatible with their linguistic world.

In recent times modern science has turned chance into a useful computational tool. For example, the basic philosophy of Monte Carlo methods used by physicists is the same as Swift's one as in this case scientists do generate random combinations of elementary events to explore complex regions of the space of events proper to a given special problem [Landau, 2002]. Physicists resort to this strategy as they are aware that very complex problems cannot be dealt with within the traditional framework of formalized and analytical theories.

As a matter of fact, in the typical case the theoretical framework can be explicitly formulated but the decisive bottleneck is due to the fact that either the theoretician lacks proper knowledge of some elements that cannot be measured (e.g. microscopic interaction forces) or the computational effort is out of reach even of the most powerful computers. To circumvent such difficulties scientists make use of adaptive strategies or computational tools using chance and selection to build up new theoretical modules or to make explicit the quantitative consequences of fully known theories. Beside Monte Carlo methods, typical examples are provided by the widespread use of neural networks and genetic algorithms [Serra, 1990; Compiani, 1996]. Such devices were originally proposed in the framework of Artificial Intelligence studies and are nowadays employed in a variety of fields ranging from bioinformatics to physics.

Paradigmatic applications were proposed to study the structure and dynamics of biological macromolecules (proteins and nucleic acids) [Compiani, 1998; Casadio, 2000; Compiani, 2004; de Sancho, 2006] as well as in the area of atmosphere physics and meteorology [Pasini, 2006].

Neural networks and genetic algorithms are artificial devices whose architectures are inspired by the cognitive strategies used by the living systems. Neural networks mimic the working principles of our brain whereas genetic algorithms reproduce the evolutionary dynamics of populations and the architecture of the immune system [Serra, 1990].

Many areas of life sciences have brought to the foreground the role of chance in several basic mechanisms. Modern evolutionary theory provided the first framework in which the randomness of a genetic mechanism was invoked as the primary engine producing novel features and ultimately driving evolution [Smith, 1975].

However Darwinism as an explanatory framework is not at all confined to the biological world. It has generated a new paradigm known as Universal Darwinism and as such it provides explanatory schemes in the most diverse theoretical frameworks [Blackmore, 2000]. In particular, science and culture are viewed as the product of a blind process of evolution. Karl Popper himself advocated the view of science as a tool of esosomatic evolution and adaptation. According to the Austrian philosopher, scientific progress is the product of a trial-and-error strategy which is similar to the one occurring in the biological world and is part and parcel of a larger evolutionary process [Popper, 1972]. Therefore, there is a process of cultural evolution beside the process of biological evolution. The hot question regards the interference taking place between these two concurring mechanisms. Bateson suggests that under present conditions it is a destructive interference [Bateson, 1997].

But even in the view of renowned scientists like Max Planck, the development of science seemed to conform to a trial-and-error scheme. He wrote that '... in order for a scientific idea to be supplanted by a new one it is necessary to wait for the physical death of its supporters' due to the reluctancy of scientists to change their own personal views. Planck's views are consistent with the so called theory of memes, recently set forth as a mechanism for the propagation of cultural items [Blackmore, 2000]. It postulates that humans are prone to compulsively replicate and spread behavioural modules or cultural and linguistic items. On top of that individuals produce tiny variations thus providing a mechanism of random change. The very essence of memetic evolution is that it takes place independently of genetic evolution and is governed by its own mechanisms. A notable difference is that the fitness (adaptive value) of the memes is usually assigned by the individuals on the basis of subjective perceptions, connected with the supposed (not necessarily real) usefulness of the memes.

This suggests that the establishment of personal convictions and opinions is ruled by a mechanism based on the intertwined dynamics of casual variation and selection. It is clear that such a role of casuality undermines the faith in the so called scientific progress intended as a purposefully directed and fully controlled enterprise. The momentum ascribed to casuality and loose control point to the role of irrational or at least non intentional factors in directing the activity of the *homo faber*.

The evolutionary paradigm has spread in other areas to the extent that not only the product of the human intellect but brain processes themselves may be studied in terms of casual change and selection. This is the very essence of the Neural Darwinism view proposed by Gerald Edelman. The units undergoing such a Darwinian mechanism are clusters of cooperating neurons and selection is carried out by reinforcing or weakening neuronal connections [Edelman, 1987, 1994].

A similar turn occurred in immunology as Jerne and Burnett showed that an inborn genetic (casual) mechanism is the essential component of the immune system. Random production of antibodies allows the organism to face the ever changing environmental conditions, albeit the organism itself has a finite temporary repertoire of antibodies. The most effective of them are positively selected in a lifelong evolutionary process taking place at molecular level [Corbellini, 1990]. On the whole we find again the synergic action of casual variation and selection that stepwise enables the organism to set up an effective immune response.

The very lesson we get from the physics of complexity is that the use of randomness (in the theory of stochastic processes) or the emergence of chance (see the case of chaotic systems) compel us to resort to explanatory strategies which require a looser control of the system than in the classical deterministic approach. Actually, in most of twentieth century physics this is accomplished on passing from a deterministic to a statistical description. However, the strategy of lowering the level of control enters a new stage in the recent developments of the physics of complex systems. In these fields of research we pass from a limited control on the quantitative results deduced from a fully formalised theory to a stage where we have only partial command on the theoretical structure of the theory itself. These developments are connected with the application of the so called artificial neural networks (ANN) and point to epistemological problems of remarkable interest.

In these cases scientists delegate to the ANN the construction of a part of the theoretical description. Actually, the ANNs can be used not merely to process data (number crunching) but also as devices that 'synthesize' an artificial (i.e. machine made) piece of the whole theory. In this sense the ANN's implement what Bateson called deutero-learning, or 'learning to learn' [Bateson, 1989]. Clear examples are to be found in the area of the ANN-based studies concerning protein folding [Compiani, 1998; Casadio, 2000; Compiani, 2004].

Whenever this occurs the piece of artificial theory is however coded in the machine circuitry in such a way that it is not accessible to the human operator. Therefore we face the challenging situation where the whole theory is composed of two heterogeneous parts. The first one is a usual theoretical tool coded in the traditional formalisms, and therefore completely explicit and controllable. The other segment is a kind of implicit, subsymbolic theory that escapes any control. These kinds of hybrid theories are not considered in Salmon's classification of scientific theories [Salmon, 1992]. Indeed one might ask whether they can be likened to a scientific explanation though, in practice, they are used as if they were equivalent to an explanation, albeit with a very peculiar structure. In any case we are dealing here with a sort of hidden rationality, due to our inability to make the hidden pieces of knowledge buried in the ANN fully explicit. In some cases, however, such hidden elements can be extracted from the ANN in order to bring them back to light [Compiani, 1998, 2004]. In the cases we regain some pieces of information coded in the inner of the ANN's structure we indirectly support the idea that the ANN has synthesized an artificial theory of the phenomenon at hand.

It must be stressed that there are no random factors in the ANNs, but the net effect is in some way similar since in the last analysis the scientist must come to grips with the very fact that he/she has only partial control on the theory and its results. This is a critical turn in which part of the theoretical work is delegated to the ANN thus introducing a kind of structural uncertainty, as far as the overall architecture of the theory is concerned.

Coming back to the evolution of culture, the main issues concerning limited control and inherent randomness are also found in current speculations on the growth of technology and scientific culture. Studies like the history of technology by Basalla and the works by Daniel Dennett depict technical and scientific development as a blind search in a kind of Design Space, with no predefined goals. This seems a typical feature of postmodernity [Blackmore, 2000] and technoscience [Latour, 1994; Lévy-Leblond, 1998]. A forerunner and bright example of this attitude is the Renaissance Italian architect Leon Battista Alberti who, as early as 1452, wrote that '... any man on earth feels urged to build and to create anything once he possesses the technical means for doing that' [*De Re Aedificatoria*, I, Prologo, I, 5]. Activity seems to be unleashed and independent as never before of any ethical value or in general any metavalue, being essentially devoted to the full exploitation of the available technical and theoretical tools.

5 Conclusions

By way of conclusion one can say that randomness and, more generally, the condition of limited rationality are pervasive elements of modern scientific and philosophic thought. This is signalled in particular by the explicit use of chance as a cognitive tool or the use of ANN's. This new attitude towards uncertainty and chance, that dates back to the beginning of the twentieth century, marks a transition from the era where the mankind was passively subjected to the effects of chance to a new era where chance and limited rationality enter routinely as main ingredients of the theoretical work (model and theory building).

In other words randomness and limited rationality are part and parcel of the set of our theoretical instruments. This confirms the early intuitions of those, like Bateson, who considered scientific language as a mix of 'loose and strict thinking' [Bateson, 1989], contradicting the neopositivistic myth of the attainability of full control of the theoretical languages, and those, as Duhem, who realized that scientific theories were comprised of an inextricable mixture of 'strong and narrow thinking' and 'broad and weak thinking' [Hesse, 1966].

Bibliography

G. Bachelard, *Il Nuovo Spirito Scientifico*, Laterza, Bari 1978.

G. Bateson, *Steps to an Ecology of Mind*, Ballantine Books, New York 1989.

G. Bateson, *Una Sacra Unità. Altri Passi Verso un'Ecologia della Mente*, Adelphi, Milano 1997.

S. Blackmore, *The Meme Machine*, Oxford University Press, Oxford 2000.

E. du Bois Reymond, *I Confini della Conoscenza della Natura*, Feltrinelli, Milano 1973.

R. Casadio, M. Compiani, P. Fariselli, I. Jacoboni and P. L. Martelli, 'Neural networks predict protein folding and structure: Artificial Intelligence faces biomolecular complexity', in SAR QSAR Environmental Research, 11, 2000, pp. 149-182.

G. Colli, *La Nascita della Filosofia*, Adelphi, Milano 1980.

M. Compiani, 'Remarks on the paradigms of connectionism', in A. Clark and P. J. R. Millican (eds.), *Connectionism, Concepts, and Folk Psychology. The Legacy of Alan Turing*, Clarendon Press, Oxford 1996, pp.45-56.

M. Compiani, P. Fariselli, P. L. Martelli, R. Casadio, 'An entropy criterion to detect minimally frustrated intermediates in native proteins', in *Proceedings of the National Academy of Sciences USA*, 95, 1998, pp. 9290-9294.

M. Compiani, E. Capriotti, R. Casadio, 'Dynamics of the minimally frustrated helices determine the hierarchical folding of small helical proteins', in *Physical Review E*, 69, 2004, pp. 051905(1-8).

G. Corbellini, *L'Evoluzione del Pensiero Immunologico*, Bollati Boringhieri, Torino 1990.

D. de Sancho and A. Rey, 'Assessment of protein folding potentials with an evolutionary method' in *J. Chemical Physics*, 125, 2006, pp. 014904(1-9).

G. Edelman, *Neural Darwinism*, Basic Books, New York 1987.

G. Edelman, *Bright Air, Brilliant Fire*, Penguin Books, Londra 1994.

E. Fox Keller, *Il Secolo del Gene*, Garzanti, Milano 2001.

M. B. Hesse, *Models and Analogies in Science*, University of Notre Dame Press, Notre Dame 1966.

M. Horkheimer and T. W. Adorno, *Lezioni di Sociologia*, Einaudi, Torino 1966.

S. A. Kauffman, *The Origins of Order*, Oxford University Press, New York 1993.

P. S. de Laplace, *Saggio Filosofico sulle Probabilità*, Theoria, Roma-Napoli 1987.

D. P. Landau and K. Bender, *A Guide to Monte Carlo Simulations in Statistical Physics,* Cambridge University Press, Cambridge, 2002.

B. Latour, *Science in Action*, Harvard University Press, Cambridge 1994.

J-M. Lévy-Leblond, *La Pietra di Paragone. La Scienza alla Prova ...*, CUEN, Napoli 1998.

Lucretius, *De Rerum Natura*, II 313-322, Rizzoli, Milano 2001.

H. Miller, *Tropico del Cancro*, Feltrinelli, Milano 1967.

J. Monod, *Il Caso e la Necessità*, Mondadori, Milano 1972.

R. Musil, *Skizze der Erkentniss des Dichters*, Summa, Hellerau 1918.

A. Pasini, (ed.), *Kyoto e Dintorni. I Cambiamenti Climatici come Problema Globale*, Franco Angeli, Milano 2006.

K. R. Popper, *Epistemologia, Razionalità e Libertà*, Armando, Roma 1972.

K. R. Popper, *Logica della Scoperta Scientifica*, Einaudi, Torino 1970 (Appendix *X).

W. Salmon, *40 Anni di Spiegazione Scientifica*, Muzzio, Padova 1992.

F. Saxl, *La Fede negli Astri*, Bollati, Torino 1985.

R. Serra, G. Zanarini, M. Andretta and M. Compiani, *Introduction to the Physics of Complex Systems*, Pergamon Press, Oxford 1986.

R. Serra, G. Zanarini, *Complex Systems and Cognitive Processes*, Springer, Berlin 1990.

M. Serres, *La Naissance de la Physique dans le Texte de Lucrèce*, Les Editions de Minuit, Paris 1977.

E. Severino, *Legge e Caso*, Adelphi, Milano 1980.

J. M. Smith, *The Theory of Evolution*, Penguin Books, Harmondsworth 1975.

C. P. Snow, *The Two Cultures and a Second Look. An Expanded Version of the Two Cultures and the Scientific Revolution*, Cambridge University Press, Cambridge 1963.

E. Zola, *La Fortune des Rougon*, Fasquelle-Le Livre de Poche, Parigi 1966.

Mario Compiani, Dipartimento di Scienze chimiche, Università di Camerino

Maurizio Dapor

In praise of chance

Abstract:

Chance has played, and continues to play, a constructive role in innumerable environments: in physics, in biology, in neurosciences, in social sciences. This paper briefly describes the crucial role that chance plays in a few important instances, concluding, in particular, that chance was fundamental in the definition of the processes that occurred during the billions of years required to achieve intelligence and consciousness.

Keywords:
chance, complexity, consciousness, random numbers

1 Introduction

In physics and biology, chance plays a very important role. In the life sciences, for instance, chance (the *mutations* that eventually lead to new species) and the law (the rules represented by the *natural selection* of the living systems suitable for survival) are present in just the amounts necessary to give rise to new biological information, and to provide enough time for trying it out.

A biological piece of information corresponds to a set of meaningful data; i.e., to symbols that are significant – representing actual knowledge – both for the whole natural world where they are brought into existence, and for the environment into which they are transferred. The genetic data are *understood* by the amino acids. They are comprehensible symbols, provided with semantic content.

Chance has played a crucial role in the identification of the language of life. Among all the possible codes that might be achieved using triplets of four distinct nucleotides associated with twenty different amino acids, it is very difficult to find a sensible and rational reason to explain the choice of our *particular* genetic code. It is interesting to note that the human genetic code, or *genome*, is very similar to that utilized by *all* living organisms on Earth. Brought into existence by the joint work of chance and natural selection, it is the only language of life we know. It is unique. In billions of years, no living systems have been developed, to our knowledge, through the use of a fundamentally different code. The basic genetic code seems, in fact, to be an example of a *frozen accident*; that is to say, something that happened and never changed.

To proceed, it is now convenient to examine the concept of *chance* from the point of view of the physical sciences, in order to establish whether or not it can be considered a *physical quantity*.

2 Percy Bridgman and the operational point of view

In physics, scientists tend to consider only quantities that can be defined in a practical way, as distinguished from a contemplative one. The so-called *operational point of view* possesses formidable descriptive potential and great conceptual wealth. To select the quantities that deserve attention and are worthy of interest constitutes a very rigorous, honest, truthful, reliable and sombre practice. The physical sciences are quantitative sciences: physicists look for relationships connecting *measurable* quantities.

This is the reason why the American physicist Percy W. Bridgman introduced the concept of *operationalism*[1]. Operationalism permits, literally, the identification of a physical quantity with the experimental procedure used to measure it. 'In general,' wrote Bridgman, 'we mean by any concept nothing more than a set of operations; *the concept is synonymous with a corresponding set of operations*.'

Let us consider, for example, the concept of *time*. Before Einstein's revolutionary Theory of Relativity, the constant flow of time was a kind of privileged entity that nothing could modify or influence. In

1 Bridgman, 1927.

Newton's universe everything changed except for the constantly flowing river of time.

Bridgman suggested that in giving the definition of time (and of any other physical quantity as well) one should avoid any vague and imprecise connotation. 'It is evident,' noticed Bridgman,

> that if we adopt this point of view toward concepts – namely, that the proper definition of a concept is not in terms of its properties but in terms of actual operations – we need run no danger of having to revise our attitude towards nature. For if experience is always described in terms of experience, there must always be correspondence between experience and our description of it, and we need never be embarrassed, as we were, in attempting to find in nature the prototype of Newton's absolute time.

In practice, physicists confer great importance upon the 'recipes' describing the processes and the measuring instruments. They are inclined, on the other hand, to consider with suspicion any concepts that are vague or difficult to measure, such as those of, for instance, intelligence, knowledge, consciousness or self-awareness. Many physicists tend – sometimes deliberately – to associate a physical quantity closely with the procedure for its measurement. Merging a physical quantity with its measurement can be fruitful and profitable, if utilized with moderation, wisdom and sensibility. However, overly strict and rigorous uses of this approach will lead to pedantic and useless distinctions.

Following a 'strong' interpretation of the concept of *operationalism*, a given distance, for example, *is* the procedure used by the physicists to establish its value. Strictly sticking to this program, we should view *as two distinct physical quantities* the distances between the macroscopic things of daily experience (measured with a rigid ruler) and the distances between the atoms in a crystalline structure (measured by X-ray diffraction).

Anyway, once lightened by overzealous applications, the idea that the physical quantities require an operative, practical definition has played (and still plays) a fundamental role in the development and successfulness of the physical sciences.

In the opinion of many physicists, chance is a *measurement* of our ignorance. On the other hand, as we have just discussed, the word 'measurement' characterizes the actual physical quantities as distinguished from other entities, not appropriately defined. Thus, because, for many physicists the concepts worthy of interest are just those that are measurable, chance seems to deserve to be considered a proper and fitting physical quantity.

Defining chance as a measure of ignorance corresponds, then, to conferring upon it the status of a physical quantity. It is a very special entity. In fact we are talking about the measurement of a subjective quantity: our ignorance. If nobody were around, chance would not be measurable.

3 Strings of numbers

Let us now discuss some concrete situations – in particular, strings of numbers. A string of numbers can be the result of either a random process or a deliberate action of selecting them. Let us imagine throwing a (perfect) coin and seeing on which side it falls. We assign the value 0 to one side ('heads') and the value 1 to the other ('tails'). Throwing the coin eight times, there is a non-zero probability of obtaining the sequence a = 10101010. Nevertheless, nobody would consider such a sequence as a random one. Its regularity (its periodicity) suggests that it has a negligible probability of being the effect of an authentic random process (such as throwing a coin). However, after a brief reflection, it becomes clear that this intuition is deceptive. Indeed, the probability of obtaining the much less regular string b = 10011100, when throwing the very same coin, is, of course, exactly the same as that of obtaining string a.

The matter appears to be very subjective and strongly related to a human interpretation of the meaning of chance: it seems so probable that a sequence as regular as a should be the result of a deliberate action, rather than of chance, that we are inclined to believe it cannot be an authentically random sequence. So, even if we recognize that string b has the same probability as string a of appearing in a sequence of coin throws; and that, on the other hand, b could have been written according to a deliberate action (indeed, *I have written it* with a deliberate action and without throwing any coin), we tend to consider *a* as regular and *b* as random. (Certainly the reader will forgive me for the somewhat relaxed and vague use of these concepts.) There is an intrinsic property of string *a*, its periodicity, that makes it special to our human eyes, even if it has the same probability as string *b* of occurring as the result of a random process.

Why does *b* seem, to us, more random than *a*? Let us group in adjacent pairs the digits of the strings *a* and *b*, starting from the first on the left. While, in the first string, only the 01 substring appears, in the second string all of the substrings, 00, 01, 10, and 11, are present. This is the reason why we are inclined to attribute greater randomness to the second string than to the first. It is clear that for longer and longer strings (i.e., the really interesting ones) such reasoning would be applied to groups of digits (substrings) of greater and greater length.

The measure of chance just proposed (an idea put forward by the mathematicians B. Singer and S. Pincus) is a little less subjective than considering chance as a measure of our ignorance or lack of information. Notice, by the way, that the last definition is identical to that of *entropy*. It is quite clear that, for very long strings, the number of *disordered* (untidy) strings of the *b*-type is much greater than the number of *ordered* (tidy) strings of the *a*-type. In authentically random processes, on the other hand, all of the strings of equal length and with the same number of zeros and ones have, of course, the same probability of being observed: as a consequence, it is much more common (and hence much more probable) to find by chance a disordered string, of the *b*-type, rather than an ordered one, of the *a*-type. In other words, there are many more untidy strings than tidy ones.

The subjectivity of the last sentence makes it rather weak. That which is messy, disordered and incomprehensible for a given person might be ordered, tidy, well-organized and meaningful for somebody else. (We would probably need the opinion of an alien from another planet to be sure of that.)

Today it seems difficult (or perhaps impossible) to say something authentically original on the issue of subjectivity. In any case, a physical quantity requiring the presence of living creatures in order for it to exist and to be meaningful is rather disturbing. I do not wish to express an opinion about a matter so distant from my knowledge and competence. I do not really know whether a tree falling in a deserted forest makes a noise; and, frankly speaking, I am not so sure that the point at issue is really worthy of interest.

4 The constructive role of chance

Chance is typically presented with a negative value, as synonymous with ignorance and/or disorder. It is, in common language, and sometimes in scientific language as well, closely associated with the antithesis of order and organization, with the exact opposite of the knowledge and predictability of phenomena. In such a way, the positive aspects of chance tend to be neglected: after all, we should not forget that innumerable *random* mutations (with the help of the strict laws of replication and selection) have put us in the agreeable condition of being able to write, read and understand one another[2].

Chance deserves praise. It has a constructive role. It brings a positive benefit. I wish to describe, now, a particular use of random sequences that has innumerable applications in physics and mathematics. Known as the Monte Carlo Method, it is a numerical procedure that allows us to solve complicated mathematical problems[3]. The name of the method is connected to its distinctive feature: it deals with long sequences of random numbers, bringing to mind the games of chance in the Monte Carlo casinos. To describe the Monte Carlo Method, I shall refer to a very simple situation.

Let us suppose we wish to calculate the area of an irregularly shaped closed surface. We begin by surrounding the curve delimiting this surface with a square of a known size. If we have a random number generator, we may now generate a great number of random points uniformly distributed inside the square. When any such point happens to belong to the surface whose area we wish to measure, we just add 1 to a variable, M. If N is the number of points generated, then as N increases, the ratio between M and N approaches the ratio between the unknown area of the surface and the known area of the square.

2 Monod, 1971.

3 Sobol, 1984.

This method applies, of course, to the case of multidimensional spaces as well. So it can be utilized to calculate the measure of 'surfaces' having any number of dimensions. In statistical mechanics it is usual to find systems with many degrees of freedom. In order to study the behaviour of these systems, scientists evaluate integrals of functions depending on a huge number of variables. An important example is represented, for example, by the calculation of the partition function for a gas with N atoms at temperature T.

It is possible to show that, for a very low value of N (say, 20), the computation of the classical partition function with any method of numerical integration other than the Monte Carlo Method would require, for a computer able to perform 10 million operations per second, a time of the order of 10^{34} times the age of the Universe[4].

Complicated physics problems involving huge numbers of particles can be treated with the Monte Carlo Method: it is possible to achieve numerical simulations of very complex physical processes, such as those represented by the interaction of an electron beam with a solid target[5].

The use of sequences of random numbers for this type of calculation is not simply useful. It is, in effect, compulsory. Chance, here, plays a constructive role: any other numerical attempt to deal with really interesting problems – i.e., those involving huge numbers of particles – could not succeed, because it would require inaccessible times of computation.

5 Chance and Complexity

The crucial issue is now the following: does chance – a measurable physical quantity used for solving real problems – play a role in making a message interesting? I use the word 'message', here, to mean any string that is useful and comprehensible to somebody. It is well known that Andrej Kolmogorov, Gregory Chaitin, and Ray Solomonoff introduced the concept of *content of algorithmic information*. The

4 Koonin, 1990.

5 Kling *et al.*, 2001; Dapor 2003.

content of algorithmic information borne by a message is defined as the length of the shortest computer code able to print the string corresponding to the message (and to end the calculation).

A very short algorithm to print a string such as 101010…10 (10 repeated *n* times, where *n* is a very large integer number) corresponds, of course, to telling the computer to write the string '10' *n* times. On the other hand, the shortest computer program able to write a random string might, in the worst case, consist of the printing command followed by the whole string. In other words, if the strings we are dealing with are very long, the highly regular sequence 101010…10 has a much lower content of algorithmic information than that of a typical random string of the same length. (Note that, while a random process might sometimes produce a highly regular string, as we have already noted, the probability of such an occurrence diminishes dramatically as the length of the string increases.)

In general, very ordered systems possess comparatively little algorithmic information. Let us consider a crystal – i.e., a regular pattern of atoms periodically repeating a three-dimensional structure. A crystal – like any other regular arrangement or disposition of things – certainly bears little algorithmic information: its content of algorithmic information is low. The content of algorithmic information of a typical random string is, on the other hand, very high (depending, of course, on the length of the string). To make the idea clearer, scientists have introduced the concept of *compressibility*. A random string is minimally compressible. In other words, it does not present any recognizable regularity to be used to reduce the length of the algorithm necessary to print it and to end the calculation. Notice, however, that there are strings which seem random but actually are not. Paul Davies has remarked that if one wrote in binary code the dates of a sequence of subsequent eclipses, the corresponding very long string of 0 and 1 would not show any regularity and would seem an authentically random string[6]. Using Newton's laws, on the other hand, one could forecast with very high accuracy the dates of the eclipses for hundreds of years to come. Hence, all the knowledge about eclipses and about the positions of the Sun, Earth, and Moon is compressible into a few

6 Davies, 2003.

mathematical equations which would barely fill a sheet of paper. All the facts concerning eclipses, concludes Davies, can be compressed in a very short algorithm and the Sun-Earth-Moon system is actually poor of information, because it shows numerous and deep regularities.

Murray Gell-Mann has studied the possibility of characterizing the complexity of adaptive systems – including living organisms – noticing, first of all, that their content of algorithmic information is neither adequate nor satisfactory to identify accurately their distinctive features and attributes[7]. Gell-Mann has introduced the concept of *effective complexity*. The effective complexity is a quantity related to the length of a computer program able to describe a concise scheme of only the regularities of the observed system. The effective complexity of a system is low when its content of algorithmic information is either very low or very high. In other words, the effective complexity is great only when the system is neither highly ordered nor highly disordered. The observer notices interesting features only if the content of algorithmic information borne by the observed system is neither too small nor too great. The two extreme cases of complete order and maximum randomness can, in fact, both be described by a very short scheme. In the first case the regularities require a very short code to be described. In the second case, by contrast, there are no regularities at all to be described.

The genome – the complete set of genetic material of a living organism – has a high content of algorithmic information. Let us organize it into a code, attributing a binary string to each basic structural unit of the double-stranded DNA helical chain as follows: 00 to adenine, 01 to cytosine, 10 to guanine, and 11 to thymine. The very long string obtained in such a way should not be very compressible. Can we actually be sure of that? After all, DNA could be similar to the eclipses of the Sun-Earth-Moon system, compressible into a small number of mathematical relationships. The claim that the knowledge codified in the DNA chain is incompressible is no more than a feeling of desire or expectation. On the other hand, I find the opposite idea – that of the existence of an equation able to compress the DNA se-

7 Gell-Mann, 1994.

quence of a *person* into a short computer code – so unbearable to suggest (let alone hope) that it must be wrong.

The effective complexity of the genome is the length of its significant part. It is, then, certainly connected to the complexity of the living organism. Yet when the human genome is compared with the genome of a primate, we are forced to conclude that the effective genetic complexity is not an adequate measure of the *biological complexity*. A more satisfactory concept, called *depth*, has been introduced by Charles Bennett. The depth of a string is related to a special average of the *elaboration times* required by a set of computer codes written in order to print the string and to complete the computation. It is worthy of interest to note that depth – in addition to effective complexity – assumes significant values only for systems neither too ordered nor too disordered.

Depth is related to the processing time necessary to give rise to a complex system. It seems to characterize, at least qualitatively, the billions of years spent to achieve intelligence and consciousness. Note that I am aware of attributing an ascending trend to evolution[8]. As such a claim is unusual within the scientific community – in which people are inclined to assume that evolution has neither an ascending nor a descending trend – I just observe that this is not a scientific issue. Any point of view should be considered legitimate in a field in which all personal beliefs and sensitivities deserve respect.

Acknowledgments

Warm thanks are due to Valentino Braitenberg (Tubinga University) and Giuseppe O. Longo (Trieste University) for their stimulating suggestions and ideas, and for inviting me to the seminar 'Teoria dell'informazione' (Rovereto, November 2003), which gave me the opportunity to discuss the issue of *chance*. I am very pleased to acknowledge Massimo Negrotti and Domenico Donnanno (University of Urbino) for inviting me to attend the VI International Conference on the Culture of the Artificial (Urbino, September 2007).

8 Dapor, 2002.

This paper is a revised version of the article: Dapor, M., Atti Acc. Rov. Agiati, a. 254, ser. VIII, vol. IV, B, 2004, pp. 37-46.

Bibliography

P. W. Bridgman, *The Logic of Modern Physics*, MacMillan, New York 1927.

P. Davies, *The Origin of Life*, Penguin Books, United Kingdom 2003.

M. Dapor, Electron-Beam Interactions with Solids. Application of the Monte Carlo Method to Electron Scattering Problems, Springer-Verlag, Heidelberg 2003.

M. Dapor, *L'intelligenza della vita*, Springer-Verlag, Milan 2002.

M. Gell-Mann, *The Quark and the Jaguar: Adventures in the Simple and the Complex*, W. H. Freeman and Co., New York 1994.

A. Kling, F. Barao, M. Nakagawa, L. Tavora and P. Vaz, Advanced Monte Carlo for Radiation in Physics, Particle Transport Simulation and Applications, Springer-Verlag, Heidelberg 2001.

S. E. Koonin and D. C. Meredith, *Computational Physics*, Addison-Wesley, Redwood City 1990.

J. Monod, *Chance and Necessity: An Essay on the Natural Philosophy of Modern Biology*, Alfred A. Knopf, New York 1971.

I. M. Sobol, *The Monte Carlo Method*, MIR, Moscow 1984.

Maurizio Dapor, Department of Engineering Materials, University of Sheffield

Masanori Funakura

Micro Rationality-Randomness and Macro Rationality in Tool Culture: Reciprocity Between Technological and Social intelligences

Abstract:

This paper is concerned with thinking about the tool culture, which is almost always in the definitely indefinable condition, as a continuous embodiment both of technological intelligence constructing the tool culture and social intelligences constructing its institutions and organizations in terms of rationality-randomness contrast, in order to consider whether accumulating tools may result in the loss of cultural unity or the increase of cultural variety in society.

Any explanation cannot escape the dependence on presuppositions assumed as most appropriate to solve a current problem within a 'possible', not strict, rational framework. As our framework is based on the following presupposition: if we start with the premise of the rational start of simple tool making to directly work on the environment at the micro level, then look at its subsequent alternate rational and random replication of tools, and, finally, consider the increasing dominance of mechanically complicated but easily usable tools, we will be able to explain the dominant trend of social self-reorientation relative with its tool culture.

Within such a framework, the following is shown: the major social institutions and organizations themselves ensuring social continuity are often the source of changes; the dynamics caused by the mutual and simultaneous interference at the micro level between technological intelligence and social intelligence in the sense of stock of knowledge can not be deterministically controlled; and the random characters of each phenomenon of tool culture may be always fixed temporarily by the convergence between two kinds of intelligences at the macro level, result in pointing to a 'possibly' consistent development.

Keywords:
tool making, tool culture, technological and social intelligences, maker and user, goal and intention, micro rationality-randomness, possibly macro rationality

1 Introduction: tool making and intelligence

Tool making has always increased our strength, speed, mobility, and comfort hand in hand with developments of new materials and enhancement of function or performance. The proposition maintained in the long history of tool making is the deterministic control of the object or environment. Anybody can work on the objects or environments using tools, direct-to-the-purpose (kitchen knife, pen, vehicle, etc.) or indirect-to-the-purpose (e-mail, word processor, computer, etc.), respectively.

Before getting in touch with the main subject, the 'tool' is defined as various types of artifacts, or articles held in the hand or manipulated by hands to do a particular job by working on objects directly or indirectly. The importance of the so-defined tool's role in social or individual life is characteristic of the present time. Japanese today is a typical example driven mainly by its market-based tool culture not less than the Unites States. More and more convenient tools become available to anyone by 'democratization' of technological intelligence. One remarkable phenomenon of the tool culture is many users' tendency of willingly buying and using newer and newer mechanical indirect-to-the-purpose tools (IT tools, in particular), though Japan is certainly proud of the tradition of high quality handicraft manufacturing. This trend that seems to occur randomly in the history of tool making seems to signal that expanding mechanically produced tools are becoming an important core of Japanese tool culture regardless of its tradition and history.

This phenomenon requires consideration of the relationship between humans, tools and society in general, with the Japanese case in mind. My concern here is that, if tool making starts rationally, at least at the micro level or locally, then, the 'possibly' rational imperative constantly promotes the tool culture into new potentials not only for enhancing physical strength but also elevating intelligence or even increasing sensitivity, and, therefore, the quantitative growth of tools may go on parallel to some profound qualitative changes.

Section one discusses that tool making starts from purpose rationality, which requires a strategy whose validity is judged by its pragmatic consequences. Section two looks at the possible randomness introduced into the tool culture by the user (and maker) at the micro levels, which can stimulate new inventions. Section three investigates the tool culture and society alternating between rational and random relationships. Finally, we conclude that any random-looking phenomenon may be rationalized by social intelligence, though the repercussions from ever-increasing change and growth of the tool culture might reach far beyond our expectation now.

2 Rational start of tool making

Any culture underlying modern society is the long-term vital process of the accumulation of both technological and social intelligences. In considering 'the origin of the idea of progress' in terms of so-to-speak 'modern' science, Rossi admits that 'knowledge' has been 'capable of an indefinite growth realized through a process of successive self-correction' as 'an attempt to control the environment and maintain the rapport of equilibrium between humans and the environment', but that the 'image of science' worthy of the name 'was absent in the so-to-speak primitive culture.'[1] Indeed, there was no 'scientific' knowledge in pre-history, but humans had technical intelligence in the sense of *tékhné* meaning 'skill' or 'art' when they started to exploit the natural environment (though technology as 'skill or craft' was later to be separated from the arts).

Nature seems to require complex reactions from species but just give back one simple and severe answer to them: survival or death. The survival motivates every species to conform to the predispositions determined by the body. For the human species, technological intelligence seems to have developed in the task of sensing and responding to such a severe natural requirement. The opposed arrangement of thumb and fingers enabled the human species to produce tools to directly work on the environment (e.g., blade, spear point, harpoon, etc).

What is necessary is just got to be that way. The coordination between manual dexterity and physical mobility due to erect bipedality

1 P. Rossi, 1995, p.45

could direct humans to resource-enriched regions or previously uninhabited zones. Indeed, humans as '*natural-born cyborgs*' was 'primed by Mother Nature to annex wave upon wave of external elements and structures as a part and parcel of their own extended minds,' and 'in quasi-evolutionary terms, the product is now poised to enter into a kind of symbiotic relationship with its biological users.'[2] Equally importantly, 'human activity leads to the creation of artifacts, shaped by human participation in natural process, ordered and guided by thought.'[3]

Considered in terms of the rationality-randomness contrast, the former (nature) preceded and the latter (human intervention) followed. Thus, the process of tool making started as a rational activity of embodying technological intelligence: practical, functional, and orderly.

Successful achievement of some purpose essentially requires the coordination of end-means and the rational strategy capable of bringing about pragmatic consequences. The strategy has passed automatically generation to generation through willing reaffirmation of their daily innovations. Then, simple tools are improved to machine other simple tools and, then, two or more simple tools were synthesized into new specific-purpose ones usable to make other tools (from a spear composed of an arrowhead plus a stick to a mobile phone or a car composed of innumerable parts). The development of tools to machine other specific production and operation came dramatically to further tool making in the field of measure, calculation, transportation, communication and so on.

In that sense, the significance of any tool at any point refers back to preceding ones by replicating their structures or mechanisms largely intact or, usually, improving in succession, and at the same time affords the chance to improve the structures or mechanisms into following ones. So the replication of previous ones can rationally develop into the production of new ones at the micro level, though changed in material, shape or performance (e.g., simple stone knives have led to iron, steel or ceramic ones). Thus, humans have continued to explore the natural environment, produce artificial environments and carefully order and

2 A. Clark, 2003, pp. 31-39.

3 D. Bohm, 1996, p. 79.

modify them. The persistent interactions between humans and their environments are the unique reason for the rational expansion of the tool 'culture', to use the term to emphasize properties related with the whole tools at the macro level, within the natural environment, involving technological and cultural differentiation (Fig. 1).

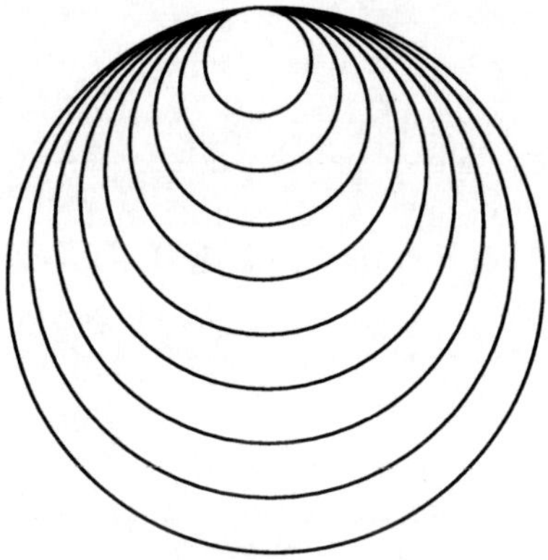

Fig. 1. Rationally Expanding Chains of Tools

In short, by including individual technological intelligence into social intelligence, humans have enlarged the tool culture as the major essential part of the artificial environment. The tool culture may seem to be the constantly expanding chains of tools successively embedded in it through repeated adherence to and improvisation of existing tools. New 'versions' can signify the more highly developed replication of preceding types, or the production of high quality new types.

3 Possible micro randomness in tool culture

Even if tool making starts rationally at the micro level and can in principle result in the macrorational tool culture, tools can cause more or less microrandom consequences in the culture in ways scarcely expected at the starting point. As any tool is nothing but a thing unless the user actually makes sense of and recognized it as useful. The purpose of the user in using a tool is closely connected with some intention. According to Searle, 'Intentionality is that property of many mental states and events by which they are directed at or about or of objects and states of affairs in the world [...] if I have an intention, it must be an intention to do something'.[4] The user uses the tool with the intention of working on the object or environment, independent of the maker's intention, in a vast complicated tool culture of designer-maker-user relationship.

The force working between the intention of the maker and the intentional behaviour of the user is only describable roughly by tracing the user's actual use of a tool as the result back to the maker's intention as the cause. Naturally, the way of using any tool is predetermined by the maker's intention of doing a particular job towards a particular purpose. Semiotically, however, the tool can be considered to be a sign as an externalized intention of the maker in that it receives its meaning through the interpretation by the user's subsequent thought or action.[5] The tool as a sign is a composite of the maker's primary future-directed intention and the user's secondary present-based action (Fig. 2).

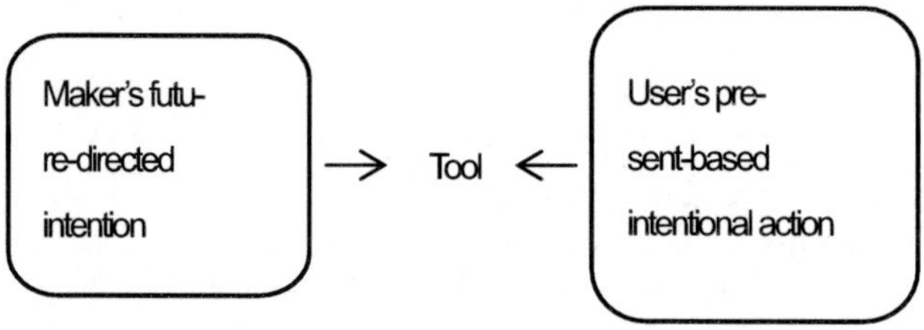

Fig. 2. Tool as Composite of Two Intentions

The user can produce an interpretation equivalent to, or more developed than the original definite one presented by the maker.

Every use of a tool corresponds to a micro rationality on the side of tool making in general, and micro rationalities with some common specified property make up a macro rationality. Seen as a single phenomenon, every tool use is intrinsically characteristic of rationality at the micro level. Each use of a specific-purpose tool is the linear process of so-to-speak personal 'microrational' intelligence.

4 J. R. Searle, 1983, p. 1.
5 C. S. Peirce, 1868.

Given just a purpose, most of the users perform by creating automatically a kind of operation program even without being conscious of the maker's intention. For example, a maker's intention of producing a tool to help the user to cut meat easily is realized by the user's intentional action to cut meat in the intrinsic way of use towards a purpose predetermined by the former. As long as this correspondence is maintained in repeated uses, even if unconsciously, the maker's intention can be satisfied just by the user. The least random class is the tool whose actual uses form an unbroken sequence.

However, any user generally uses a tool in the mental state resulting from their hitherto course of intelligence deriving from interactions with other tools and users in the tool culture. Some users may realize the 'toolness' in their individual mode of use in certain extrinsic ways by investigating what they can do with the tool in hand. The user may even perform by creating and generalizing some operation program irrelevant to the original intention in the process. So it can be easily inferred that any tool can be turned to other uses. For a simple but fundamental example, a stone knife was first made towards the specific purpose of hunting animals and cutting the meat of some captured animals. Then, it was applied, by chance or intentionally, to another use of attacking other tribes: the tool was changed into a weapon. After that, it was in turn developed into a more general use of cutting anything. It is never a logical jump that any tool can be used in extrinsic ways (automobile for killing others or sleeping, mobile phone for spreading misinformation, etc). This example shows how to use the tool depends on the user's intention. High possible randomness may be found in the tools in which the functions are not all of the same kind and their actual uses don't possibly form an unbroken sequence. Tools usable in the most extrinsic way cause randomness in the tool culture (Fig. 3).

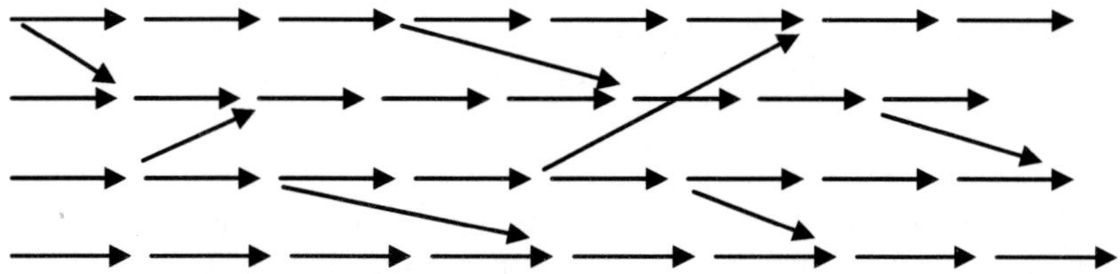

Fig. 3. Possibly Micro Rationality and Randomness

When such incoherence occurs between the user's intentional action and the maker's expectation, the user will regard their own action as rational, only if they can reach their desired purpose. It is because what they do with full awareness of what they are doing can be considered intentional and intention creates a framework of rationality within which the action plans can be followed and controlled, though such plans might look possibly random from outside of it.

To summarize, many simple-mechanism specific-purpose tools survive in the tradition of linear tool-making. They continue to be the same in essence by being rationally maintained and developed. On the other hand, several specific-purpose tools have been synthesized into more and more complicated tools, and the tools synthesized into those new ones turn into their parts, being no longer independent tools. They continue to change themselves randomly, though designed rationally at their micro level. Many phenomena of tool making and using show two alternate states: the ordinary state in which all users use tools in intrinsic ways predetermined by the maker, though they can enhance their intelligence and skill by controlling the tools in the predetermined scope of use; and the deviant state in which some users use and control tools in extrinsic ways to enhance the function or performance of the tools as well as their intelligence and skill outside the predetermined scope of use. The transition from one to the other states may occur due to a fluctuation, intentionally caused by new technical discovery, invention or improvement on the maker side or intentionally or accidentally by some chance discovery of ways of customizing, elaborating or abusing on the users' side.

4 Possible macro-rationality of tool-based society

Many tools have pursued their wider use by what may be called 'democratization' of technology. In the increasingly large-scaled and complex systems of production, transportation and use of tools, the increasingly more complicated tools are randomly inserted into the tool culture. Unlike the tools whose ways of use are predetermined but controllable by their users, the tools whose scopes of use are also pre-programmed and beyond their users' control can have more influential effects on society. Hence the necessity of considering a social problem of such indirect-to-the-purpose tools occupying the core of the tool culture.

In order to consider the substance and ontological logic of the ever on-going tool culture, look at Japanese increasingly technological society as a typical example, in which social and technological systems seems to be organically fused into one. Roughly, since the latter half of the 20th century, many tools have been popular for personal uses. Once tools in general were mainly a simple-mechanism direct means of satisfying everyday life conditions. Now, technology continues to bring 'at our fingertips a range of facile skills each of which we would earlier have had to rely on others to provide. Now that the "skills" are universal, easy to acquire and learned from scratch, they come without roots. When we venture out into the world we have no background.'[6] This point may be understood so easily by taking the mobile phone as a typical example of complicated indirect means.

Now more mobile such as mobile phone, computer, PDA tools are being personalized by the Japanese. For personalization of mobile tools, the watch or pocket radio may be thought of as a forerunner of such a personalized tool. But they can not stand comparison with the IT mobile tools, in particular the I-Mode (wireless internet access) mobile phone with the functions of e-mailing, calling, playing games, listening to music, taking photos, watching TV, and so on. It gives the user a literally personal significance as part of their life history like the automobile to the American.

6 A. Hannay, 2005, p. 80.

Indeed, the mobile phone seems possibly to cause radical changes in almost every aspect of life and in the relationship in particular between society and its members. This tool can make its users feel as if living a more personal and liberated life than ever before. More importantly from the perspective of human-tool relationship, however, the mobile phone is a typical example of indirect-to-the-purpose tools. The user's job is completely entrusted to the maker's intention in the scope of the tool's function.

As tools become more synthetic in mechanism, they go further apart from their users. The performance of most exclusive tools depend on their users' skilful management. They are controlled by their users when directly working on objects or environments. Those tools allow their users to enhance the potentials as tools by exerting their ingenuity. In other words, their skill or intelligence and tools' potentials go hand in hand in an ever deeper relationship. To such skilful users, their tools are the other self, or even life itself.

On the other hand, the IT tools, in particular, the mobile phone or computer, generally require really little intelligence and skill from their ordinary users, even though the personalization of them may make them believe in activating them. Certainly, in the sense that the user does not know in detail the mechanisms of tools or parts synthesized into one tool but can operate it as a whole, the many-purpose IT tools used in the indirect environment. The mechanism of those tools are the media through which the users indirectly reach the purpose and which, therefore, they can not control in details while the specific-purpose tools are directly operated and, therefore, controlled by the users in the direct environment. They seem to operate tools as independent agents, but they behave mainly through the uniformed programs, or the makers' intelligence, as if by advance arrangement. The users simply enter into an increasingly surface relationship with those tools.

If this tool phenomenon poses the problem of a major possible change in the meaning and form of social intelligence, what is crucial is that the extensive development of high technological intelligence does not grow into the dominant social intelligence, or is not 'democratized'. Because the rapid development of IT tools and their systems are too difficult for ordinary udders to understand but too easy to use, they are controlled by a small group's intelligence. Here is a good suggestion from Vattimo: the 'ideal of emancipation' from the threat of the control of power in the society of media should be not 'on the perfect consciousness of those who see how the things are', but 'on the proper basis, rather oscillation, plurality, and, in short, the erosion of its own principle of reality'. It is suggested that we choose 'this oscillation as liberty' and 'this experience of the oscillation of the post-modern world as *chance* of a new mode of being (probably, finally) human'.[7] Indeed, the idea of oscillations instead of forcefully rational intervention in the event may be acceptable to Japanese culture, which can be characterized as drawing a hazy line of demarcation between rationality and randomness.

Paradoxically, Japanese fundamentally resistant-to-change society seems to identify with its conscious logic-based American rationality and institutionalize it in the tool culture, and displace the unconscious randomness on which its conscious identity is built. If Japanese society comprises both conscious rationality and unconscious randomness at the macro level, such a layered combination will neither make a social context acting as a coherent whole in some sense in which tools are used nor place social effectual pressure on the interpenetration of randomness and rationality found everywhere in the shifting patterns of humans' intellectual activities themselves.

If any tool-based society like Japanese one attempts to re-adjust itself to the situation by appealing to such instrumental rationality, which justifies accepting the reality as it is, it will fall into self-contradiction, because any society changes consistently driven by its own dynamism of continuous complicated interactions among technology and society including individual users in a complex and random system. New discovery are often due to chance or random conception, which is born from the viewpoint opposite to rational one. Naturally, the two conceptions come into conflict with each other at the outset. If some pattern appears that indicates a kind of rationality within randomness, it will surely lead creativity to a new direction. The details that seem insignificant now may develop into important causal factors in future, while the trend that seems obvious and unavoidable now may get derailed for the reason currently considered insignificant. In short, for society to be

7 G. Vattimo, 1989, pp. 19-20.

mobile and active, we must leave its feedback control to its own autonomous operation to a certain extent. The whole event first seems merely to randomly change, but the movement towards coherence seems to be innate to any culture, though it produces incoherent values which often muddle us in the way.

5 Conclusion

The major social institutions and organizations that give concrete expression to continuity in our society are often themselves the source of changes. because we are part of the system both of technological intelligences embodied in the tool culture and social intelligence embodied in the institutions and organizations In the high tech society, technological intelligence replication must be seen as possessing its own dynamic promoting social intelligence reorientation. So we cannot deterministically control the dynamic caused by the mutual and simultaneous interference at the micro level between technological intelligence and social one. The convergence on each occasion between them should fix the random character of each tool phenomenon at the macro level and display a consistent tendency of society towards possibly macrorational self-reorientation as a whole. The tool-based society is always a provisional, never perfected system because technological intelligence is an ongoing process of transformation and generation.

Bibliography

P. Rossi, 'Sulle origini dell'idea di progresso', in *Naufragi senza Spettatore*: *L'idea di progresso*, Il Mulino, Bologna 1995.

A. Clark, *Natural-Born Cyborgs: Minds, Technologies, and the Future of Human Intelligence*, Oxford University Press, Oxford 2003.

D. Bohm, 'The Art of Perceiving Movement', in L. Nichol (ed.), *On Creativity*, Routledge, London and New York 1996.

J. R. Searle, *Intentionality*, Cambridge University Press, Cambridge 1983.

C. S. Peirce, 'Some Consequences of Four Incapacities', *CP* 2 pp. 211-42, 2.303, 2.92. 10.

A. Hannay, *On the Public*, Routledge, London and New York 2005.

G. Vattimo, *La società trasparente*, Garzanti, Milano 1989.

Masanori Funakura, Division of Language and Culture Studies, Tokyo University of Agriculture and Technology

José M. Galván

Creation and Casuality: The Case in Christian Theological Anthropology

Abstract:

Overcoming the different forms of determinism present in some natural religions, the Judeo-Christian revelation permits an anthropologically positive ethical vision of chance. Chance is referred, as in Aristotelian philosophy, to the interlacing of two causal lines, one of which is unable to control the other. Therefore that which happens 'by chance' does not negate either divine omnipotence or the fact that God created everything for love, but has the effect of pushing mankind towards an ever greater understanding of causes, thus making it possible for history to be seen as a process of growth guided by human liberty.

Keywords:
chance, natural religion, Judeo-Christian revelation, creation, history, freedom, knowledge

The object of this contribution is that of shedding light on some aspects of the revealed Christian doctrine that underline the inclusion of the case in a positive vision of man and his fundamental ethical dimension.

It is not difficult to find in the area of the religions, the negation of the case. Absolute determinism that usually characterizes many points of view of the divinity and its operation, as well as the anthropomorphization that is sometimes excessive for both, leads logically to the failure to recognize the possibility of an effect that is not linked in a 'personal' manner to the divine causality. In the absolute sense, this vision is shared by Catholic theology[1], but with the emphasis that this negation has to do fundamentally with the need to affirm the immutability of the divine decrees, and likewise remain clear that the divine government of the world does not impose any necessity on the created wills, for which the free personal actions are not wholly controllable or foreseeable; as a result of this unforeseeability, the case is therefore inherent to a system where freedom is in play. However, it is a 'non-absolute' case: God plays the dice, but always wins.[2]

Every determinism is a negation of the case. The religion that, like the helenistic one, admits instead a subordination of the divinity to a superior impersonal force, a sort of *kratos* of the cosmos, tend to give the case, to that which is fortuituos, a preeminent role. Maybe because some authors like Epicurus or Lucretius, following Democritus in a general way, have spoken of fortuitous encounters of atoms; but have not given reason of their existence and much less of the movement of these, they have not been capable of explaining the uniformity of the effects that follow them.

Modern scientism also, just like the determinism of Democritus, finds in the concept of the case, an important support. To apply for instance, this concept to the origin of being and to its evolution permits anthropologically a total eradication of a past that transcends us, to find the comprehension of itself in the exclusive area of positive science. This occurs many times and paradoxically also when the case is negated: Newton, Laplace, Einstein, Popper, in diverse proportions and tonalities, have sustained the idea that the case corresponds to a lack of knowledge of the laws that determine an event (Einstein: 'God does not play dice'). For them, to admit the case is only a manifestation of ignorance, a challenge to look for and discover the 'unknown cause'. This would seem at first sight a subsequent negation of the

1 The revealed doctrine about God negates the 'fact' or whatever other force superior or parallel to God (cfr Agostino d'Ippona, *De Civitate Dei* 5,1; Tommaso d'Aquino: 'nec quidquam accidit casu in ordine universi: casus simpliciter dictus omnino impossibilis est...', *Summa Theologiae* I q. 103 a.7; cfr. also q.116 a.1).

2 Behind this apparently paradoxical affirmation we find the correlation between the affirmation of human freedom and of the personal divine omnipotence that founds it, a constant argument in the history of theological anthropology, that obviously is not possible to undertake in this occasion.

case, and a contradiction of the discourse, but if one takes into consideration that which empirical science attains, it is not a reduction of the case to that which would be aristotelically called the cause, for 'cause' in the area of positive science does not usually have an ethiological or teleological connotation, but fundamentally descriptive of giving oneself of a determinate process; in the final analysis, 'cause' would continue to be a 'case'.

If the concept of nature is understood only in the area of technoscience that has dominated a great part of the past century, it would be very difficult to avoid the temptation of reducing man as well to that area, reducing likewise freedom to a determination of physics.[3] Given the impossibility of this reduction (because the person always tends to emerge, if looked for, in a second moment) of putting man totally out of the area of nature and its laws: the activities of the human spirit are put outside all law, and it is very difficult, if not impossible, to found an order of ethical reference.

This causes a radical sense of distrust in the encounters between persons, not capable of being dominated by the laws of empirical science and above all, very unreliable. The pretence of the scientific paradigm of dominion will be the absolutization of technique, seen as the unique way of redemption before the consciousness of the imperfection of humanity: the person auto-emargines himself. This points

3 This danger, and its solution, are emphasized on by John Paul II from the beginning of his pontificate: 'se la scienza è intesa essenzialmente come "un fatto tecnico", allora la si può concepire come ricerca di quei processi che conducono a un successo di tipo tecnico. Come "conoscenza" ha valore quindi ciò che conduce al successo. Il mondo, a livello di dato scientifico, diviene un semplice complesso di fenomeni manipolabili, l'oggetto della scienza una connessione funzionale, che viene analizzata soltanto in base alla sua funzionalità. Una tale scienza può concepirsi soltanto come pura funzione. Il concetto di verità diventa quindi superfluo, anzi talvolta viene esplicitamente rifiutato. La stessa ragione appare, in definitiva, come semplice funzione o come strumento di un essere che trova il senso della sua esistenza fuori della conoscenza o della scienza, nel migliore dei casi nella vita soltanto (...) Una scienza libera e asservita unicamente alla verità non si lascia ridurre al modello del funzionalismo o ad altro del genere, che limiti l'ambito conoscitivo della razionalità scientifica. La scienza deve essere aperta, anzi anche multiforme'. *Discorso,* Koeln 15.11.1980, Insegnamenti di Giovanni Paolo II, vol. III/2 (1980), 1200-1211.

towards a cultural system of sacralization of the immanence that, not being able to give reason to the imperfection of man, pretends to reduce the history to a process in which man every time_accounts for less and less, in such a manner that avoids the immeasurable and unforeseeable factor of the risk that comes from the person. So many prophets of the crisis of modernity, like Nietzsche or Dostoevsky, have announced that humanity had taken a path in which there wouldn't be any more room for freedom. As the Great Inquisitor emphasizes, freedom is proper to the last thing that man likes. In the end, the case does not exist, only the ignorance of the laws that permit the technical dominion of every intra-cosmical process. Man frees himself from the case, but he auto-condemns himself to not having any other sense other than an indefinite and a-teleological progress without knowing where its end lies. Man becomes a tireless ant.

It would seem, in fact, that it is not possible to speak of the case without speaking of freedom, of finished operation.

Classical metaphysics shares in a great part the initial negation of the absolute case, now that every effect has to have a cause, and the primary cause cannot be strange to any of the secondary causes: the divine causality exercises itself in the operations of every creature, through the so-called *concursus,* the foundation of which any operation of a creature requiring a minimal actualization of its own good, implies an immediate divine cooperation: *'agere nihil alius est quam communicare illud, per quod agens est actu secundum quod est possibile'*.[4]

According to this classical vision, the case is a word that manifests the accidental and unforeseen coincidence of the natural effect of a cause with the natural effect of another cause not dominated by the first. In this sense, it can be said that the case exists truly in the nature in the objective mode, not as an exception to a physical law, but as a purely fortuitous and unforeseeable event, that can be explained from the fact that in the nature there are diverse teleologically organized systems that over-place themselves, and the case comes to be when an event or a subject, pertains to a superior teleological system and in a given measure dependent on the inferior system, that it cannot control,

4 Thomas Aquinas, *De Potentia* q.2 a.1.

or that is not wholly controlled by the superior system. Like this it is possible to place into agreement the metaphysical research (nothing occurs without a cause) with the fact that the case exists truly as a consequence of an interaction of effects between at least two causal systems not mutually reducible. This way it would not be contradictory to affirm the real existence of the case and therefore of its real role in the becoming, with the affirmation of a Prime Cause, for which the case doesn't exist, in as much as it is not fortuitous or unforeseeable.

The simplest example is that of finding 'by chance' a 500 Euro note on the path when I go to work: the action made by the operational system is not directed towards the finding of the money, neither is it capable of controlling the system by which that money has ended up on the road. The positive effect (would be also negative, if instead of a note of money, one finds a hole, or a roof tile falls on my head) is properly called 'a case'.

Maybe the most common (but not the only) form by which a superior teleological system (human) is not capable of controlling that which is inferior is ignorance. This way, the case manifests many times our incapacity to foresee the convergent effects of diverse natural activities (and in this sense, it can be said that it does not exist in what refers to God).

But also, there are other factors that may increase the incidence of the case in the measure in which they make difficult or impossible the control of the inferior system. For example, the weakness of the biological cybernetic systems may cause the incidence of physiochemical factors of inferior systems to occur with higher frequency and intensity: this way the condition of sickness of an organic being increases the incidence of negative causal factors in the measure in which its cybernetic processes are deteriorated: to the sick organism, many more 'incidences' occur. Human reason may also interweave intentionally teleological systems that are not controllable so as to obtain the desired effects as unexpected, as in the game of gambling or in determinate artistic manifestations. This same 'production of the case' and its subsequent rational management with statistical methods or other systems of research ma become enormously rich for the human being and for his development.

Anyway, that which is produced by case presupposes the causes that operate in a determined and foreseen manner, in relation to those that may be called fortuitous; in fact, the case may be understood only in reference to the presence of an operation that has an end, teleologically determined, that generally is proper of the human being. The case does not exist if the cause does not exist; the case wouldn't exist if everything was causal. Nevertheless, being true that the case cannot give reason of the natural order, general and constant, that eventually we can discover in the reality that we are in, it is clear that its presence frees the cosmos from the epistemological poverty of the deterministic laplacian universe. Human history, for example, progresses thanks to the interaction between the human generic finality, events controlled by man, ignorance, the case. Human history is the mixture of the case, freedom, physical necessity, human initiative. Many progresses have taken place due to the case, maybe also in the evolution of life, but always in relation to a superior teleological system that gives it meaning.

The Hebraic-Christian idea of creation shares in a great part this positive vision of the case, and gives reason of existence and interaction of diverse teleological systems, in the first place, affirming that creation wouldn't have sense if it were not understood as an interpersonal event. In other words, God would not have created a non-personal creation, or a personal creation that is not called to dialogue with him.

Sacred Scripture, in fact, insists in presenting a God open to the world, open to a man with whom he would like to establish a communion of eternal life that technically is called *salvation*. But with the same insistence there is emphasis on the fact that God doesn't have any need of that communal dimension with the creature: He is in himself the fullness of life, and from that omnipotent fullness springs out the same creature he loves. In the Old Testament, this fullness of life, necessarily confessed but not yet revealed, is called Sanctity *(qadosh)'*, in the New Testament *qadosh* is revealed as Tri-Sanctity of intra-divine personal life: Father and Son and Holy Spirit. God is therefore communion in himself, and for this reason can be communion towards the world.

If a beloved creature exists, therefore, it is in the interior of that love that God is in his intra-trinitarian life. Otherwise, the communal reality *ad extra* would put in crisis the divine transcendence. The being of the creature is returned, therefore to that which constitutes the same Trinitarian life as *communio*: *the eternal processions.* In the words of Thomas Aquinas,

> God knowing himself, conceives in the intimacy of himself the Word and this also is the idea in which he knows all the things. From this conception of the Word, He proceeds to the love of all the things and of himself… when this PATH is closed, nothing may be added to it, in this manner there cannot be a third procession in the immanence of the divine nature, but only the procession of the creatures outside God.[5]

Dante would say more or less the same thing in less words: 'Both that which never dies and that which dies are only the reflected light of that Idea which our Sire, with Love, begets.'[6]

In effect, all divine procession *ad extra* (creation) is referred to the eternal processions of the knowledge and intra-divine love, and it doesn't have any other sense other than that of establishing a diverse reality from the divine essence, a no-God in contrast to God. Then, also the affirmation of the absolute divine omnipotence leads us to hypothesize the possibility of a merely material creation, and one cannot see the reason why God would have put in the being of a no-God, known in his Word and loved in his Love, not capable of establishing a personal communion with Him. 'Person' is a no-God called to love God. Creation (*ad extra* procession) guarantees that alterity, necessary for that love to e, that the Trinitarian processions guarantee in the interior of that divinity.

Thomas Aquinas has in fact insisted, that properly because of the personal character of creation, in the compatibility between the case and the divine Providence, dedicating to the whole theme a question of the *Summa contra Gentiles*[7]. The fundamental idea is that the per-

5 Thomas Aquinas, *De potentia,* 9, 9.

6 *Divina Commedia, Paradiso* canto XIII: 'Ciò che non more e ciò che può morire non è se non splendor di quella idea che partorisce, amando, il nostro Sire'.

7 'Contra rationem providentiae esset si res providentiae subiectae non agerent propoter finem: cum providentiae sit omnia ordinare in finem. Esset etiam contra pre-

sonal dimension of creation, that requires the fully free operation of the Primary Cause, surpasses all radical determinism and frees the creature of being teleologically abandoned.

The concept of creation includes the distinction between the creator and the creature, between God and the world; in this sense one can distinguish them from concepts like emanationism or pantheism. The world, diverse from God but at the same time in dependence on Him, is then considered from the point of view of no-absolute, and this permits the illumination of the meaning and the avoidance of the relative autonomy of man in his own actions, or at least, impedes absolutely the deterministic interpretation or the teleological lack of the person: if there exists a being that creates then the determinating *kratos* of the cosmos is not inevitable. In the system not only does divine freedom exist, but also that of the its image-creature: man.[8] This created freedom is placed, by the same creatory will[9], under a temporary

fectionem universi si nulla res corruptibilis esset, nec aliqua virtus deficere potens, ut ex supra dictis patet. Ex hoc autem quod aliquod agens propter finem deficit ab eo quod intendit, sequitur aliqua casu contingere. Esset igitur contra rationem providentiae, et perfectionis rerum, si non essent aliqua causalia' *Summa contra Gentiles* III, 74. Cfr. also P.T. Geach, *Providence and Evil,* Cambridge University Press, 1977; D.J. Bartholomew, *God of Chance*, SCM Press 1987.

8 The following text of the Aquinate shows how the case accounts in its own system as a counterpoint of personal freedom: 'Nihil potest esse homini sufficiens causa peccati, quod est spiritualis ruina, nisi propria voluntas. Et ideo dicta vel facta alterius hominis possunt esse solum causa imperfecta, aliqualiter inducens ad ruinam. Et propter hoc non dicitur, dans causam ruinae, sed, dans occasionem, quod significat causam imperfectam, et non semper causam per accidens. Et tamen nihil prohibet in quibusdam definitionibus poni id quod est per accidens, quia id quod est secundum accidens uni potest per se alteri convenire, sicut in definitione fortunae ponitur causa per accidens, in II Physic.', Summa Theologiae II^{a}-IIae q. 43 a. 1 ad 3.

9 The idea of creation consist fundamentally in affirming that the cosmos and man have had their own origin in a divine action that puts in act all being of reality; in this sense the classical definition *'productio ex nihilo'* shows well the absence of a material cause that brings it about. The scholastic extensions of this definition (*'productio alicuius rei secundum suam totam substantiam', 'productio rei ex nihilo sui et subiecti'*), even though they make the concept more precise, tend to favour a less cosmological conception of creation, for which must be adequately inserted in a personalistic ambit in which there may be present the character of intellectual agent that is included in the affirmation of the divine causality.

and spacial (material) condition by which its state is that of a succesivity of being that is given in relation to a material cosmos which in a sense is transcended from the person and which in another sense limits it. In effect, one can affirm that man in time is not objectivistically placed under the need to move from the successive instances in an automatic or determined way, but becomes the permanent and active subject of his successivity. The dynamic condition of the human being, therefore is this transcendence over the material dimension in which he is constituted. Transcendence that manifests itself not only in being superior to the rest of the material creation, but also in going farther.[10]

Man, therefore, has a certain dominion over his temporality, that permits him to be creator of himself and creator of new relations with other men and with the rest of material nature. This capacity of dominion, anyhow, finds itself limited by the very material condition of man. On the one hand the creativity of freedom acts truly, opening always the subsequent moment to a horizon of a new possibility of being; but the same fact of existence in the next moment is by itself a manifestation of the natural limitation of the being of man, that does not possess in fullness – at least not with that fullness that the concept of freedom could suppose – of its being: only the eternity of God is *tota simul et perfecta possessio* (Boethius). It is necessary to affirm without opposition these two truths of human nature: in the first place, dealing with a permanent nature that conditions the being of man as a special-temporal being; and in the second place that this is a free nature that makes of the human temporality a true historical process, of which man is subject and author.

10 'Now, man is not wrong when he regards himself as superior to bodily concerns, and as more than a speck of nature or a nameless constituent of the city of man. For by his interior qualities he outstrips the whole sum of mere things. He plunges into the depths of reality whenever he enters into his own heart; God, Who probes the heart, awaits him there; there he discerns his proper destiny beneath the eyes of God. Thus, when he recognizes in himself a spiritual and immortal soul, he is not being mocked by a fantasy born only of physical or social influences, but is rather laying hold of the proper truth of the matter.' Vatican Council II, Cost. *Gaudium et spes*, 14b

History is not a necessary process: God has willed to take to risk confiding it to the free nature of the personal creatures. This, on the part of God, opens for us the mystery with which He, prime author of history, moves in this history to the free creature inasmuch as the creature is free. On man's part, there is emphasis on the tension towards the fulfillment of his proper state in terms desired by God, as a radical role responsibility that gives to the temporal existence a profound moral meaning, and makes it become every movement of the human temporality an occasion of a true encounter with Christ.

What has been said so far permits an ethical evaluation of the case, o even more necessarily, the interaction between human operation and the case. If the ethical dimension is truly an intrinsic element and constitutive of the human being, and not a sort of convenient addition, the ethical discourse cannot maintain in every moment a strong link with the question of man and himself. However, the question about himself cannot be posed without the question of the cosmos in which man is inserted and in the encounters of which it is interpreted, be it like the sick cot of a system that threatens it, or be it perceiving in a manner more or less evident its peculiar role in the interior of the same system. The question over the origin of man and the cosmos, prime foundation of the possibility of the interpretation of its proper role, has always been present in the culture, and the central orientation of the operation towards themselves, towards the others, towards the reality as a whole, has been for humanity a result of this response. In the attempts to give consistency to the proper being and to the being of the world, the question over the origin of all the things has played always a fundamental role, above all in the religious area.

In order to continue this discourse, it is necessary to determine, at least in a general way, the nature of ethics. We hold that the concept most used as the basis is the concept of good: an ethical behaviour will be that which looks for the good of man in himself, in his relation with the others, in his being in the world; and any human person that retains that it is convenient to place himself under the rules of ethical behaviour, will agree with this, that it is a good for man and it causes happiness. In fact, classically there is a tendency to unite ethics to the search for happiness, that is found in the possession of good. However, what is the good of man? It is not possible to give an immediate answer of the objective type to this question, because man takes the good primarily just through the experience of happiness.

The most immediate of the human experiences consists in the fact that all that is known as good is capable of causing a desire of the will and an affective reaction; from this proceeds the connotation of objects as important for the person. To this desire and to this connotation of importance, follows the activity of reason that looks for the confirmation that that which is subjectively satisfactory corresponds to the real value of the object contemplated and not to its simple capacity to produce pleasurable or fulfilling reactions.

The value of the object is determined as a result of its orientation, more or less intense, to the full realization of the human life as a happy life; due to this, it is possible to establish the finality in our actions, and subordinate the partial ends to one another so that the actions can be oriented towards the most general good. Ethics has as its role, to orient the human choices towards this full destination: that of finding happiness in that which objectively adds value and not only in what is subjectively satisfactory, and to make it in a such a manner that the inferior ends are subjected to the superior ones.

The interaction of man with the others or with nature is not simply a consequence of a need to fulfill his own necessities, but rather of the intention of improving his own position and his own nature. In this sense, we can say that man not only has needs, but also is capable of creating new ones, achieving in this way a betterment of his relation with his surroundings and with the others. But this is not all, man also has the characteristic of bettering himself: he is doted with a plasticity that is practically unlimited, and is capable of defining by himself new roles or objectives, for which he can achieve modifying himself with the acquisition of new operative capacities, be they spiritual or corporeal.

The specificity of a human act is found in its own orientation towards the end and it is from the quality of this orientation that the same act obtains its ethical dimension. In this sense, we can say that the specificity of the human act is found in its free finalization towards the good. The problem occurs when this is placed in the interaction between the human operation determined towards the end with the condition of man as immersed in a system that, though teleologically inferior, is capable of influencing in a decisive manner over its own being irrespective of its own intentional capacity: this is what we have said that it occurs not intentionally, but 'by case'. The realization of the finalistic growth of man includes the fact of the need to interact with the matter in its own operation so as to 'auto-determine' itself to an end. The great ethical question is found, therefore, in the relation between the two autonomous systems that are however mutually interactive: that of the human teleological and that which is 'casual'.

According to Aristotle, it is classically called *praxis* the action that is characterized by the determinating presence of the end in the constitution of the action, and this is the type of action that is properly speaking human. To the action in which the end does not form part of the constitution of the action, but rather is found at the end of this latter one, is known as *kinesis*, that is proper to the agents that cause. In the *praxis* the end is precedent and external to the action and is determined by freedom, whereas in *kinesis* the end is immanent to the same action. A kinetic action may become praxical if it is assumed from the operating principle with the capacity to determine the end, whereas a praxical action can become kinetic when the operating principle with the capacity to determine the end renounces this possibility.

The ethical tension of the person is manifested in his efforts to convert, as and where this is possible, into *praxis* the *kinesis* of the case. Obviously this will not always be possible for the single individual, but if we consider the common subject that is humanity, I retain that it could be affirmed that in every case, that which comes 'by case', in regard to the presence of the cause that is not controllable by human praxis, may be constituted by itself as a call and a full role of ethical significance to search for a greater control of the causal event. The single individuals can only suffer (and even die) before the causal potency of an earthquake or of a simple roof tile that falls on the head; however this same facts hold for humanity precise ethical tasks, from the attempts of statistical systemization of the causal events, to the discovery of the unknown causes up to the moment, to the realization of every means of prevention or defense that the technical capacity of man may avail.

Many times the case is manifested in life as a radical expression of the mystery of man and as an enigma that requires necessarily an answer from outside; for man it is experienced as a reality that should not have been for us. The human effort to control the case (even that which is positive), instead 'should be there' for us. In fact, the personal being in a material cosmos – knowledge that constitutes the core of secularity – requires this effort as a manifestation of the specific role of man in the world.

In this sense, a creation in which the case is present becomes the occasion of growth and creativity on the part of the creature that, for being free and personal, gives reason to the existence of the case. The theorical unhappiness that indetermination of the case produces (even when it involves 'positive cases', the fact that they occur 'by case' and causing incommodity) may be converted by man into a tendency towards a greater plenitude of his own being that, not determined by the pre-constituted virtuality, auto-determines itself interacting with its own material nature and that of the cosmos in which the case is referred to a personal foundation that is transcendent that gives reason to the hope that illumines the end.

I think that the crisis of modernity started when hope, that is the reason for which man walks through history, began to be confused with an empty anthropological optimism that becomes blind because it has not had the vision of the objective. Sisyphus climbs the mountain only in the myth: in the real life, Sisyphus halts, and abandons himself to the case. In this sense, the postmodern rationality, already conscious of not being capable of maintaining that paradigm of immanent dominion of reality and needful of rediscovering a true relational link with this latter, is able to obtain help from the contents of the biblical message about man and about the world, introducing the unthinkable of his material condition in the light of his transcendent calling.

José M.Galván, Pontificia Università della Santa Croce, Roma

Giuseppe Lanzavecchia

Chance: a law of nature or a cultural construction?

Abstract:

Some aspects of reality seem due to chance, while regularities shown by nature seem to exclude it. Only science can provide an answer to the apparent dilemma, and get along without intuition and reasoning. Science has many tools to do just that, but so far it has reached no firm conclusions, even if has allowed us to explore the problem more deeply, and to clarify the concept of chance.

Keywords:
chance, intuition, reasoning, fractals, serendipity, quantum mechanics, Laplace, Heisenberg, Schrödinger, Gödel

1 Can or must we believe in chance?

Is Wolfgang von Goethe right to say, in his *Italian Journey*, 'Ego is a bottle filled at random by what overturns on it', or is every event perfectly prearranged in a rigorous Laplacian construction that we do not know how to read?

Arthur Schopenhauer suggests that 'destiny shuffles the cards and we play'. Camillo Benso, count of Cavour, declares that 'every plan, every project, is useless; all depends on chance'. Jorge Luis Borges tells us: 'I know about a dizzy country where lottery is the primary part of reality'. Schopenhauer adds, in *The World as Will and Representation*, that

> nature speaks thus: 'The individual is nothing and less than nothing. I destroy millions of individuals for fun and recreation every day. I leave their destiny

> in the hands of the more moody and capricious of my sons, Chance, who pursues them as he pleases. The individual is nothing'.

Eugenio Montale, in *The History*, recalls that

> History doesn't come untied as an unbroken chain of rings. In any case, many rings don't hold. History does not hold the before and the afterwards; nothing that is in it grumbles at low flame. History is not made by those who think it, and not even by those who don't know it. History does not clear a way, it persists, it hates little by little [...].

In the 'Discourse on the Method' René Descartes writes: 'my third principle was always to beat myself rather than fortune, and to modify my wishes rather than to change the order of the things in the world'. For Anatole France 'chance is perhaps the pseudonym of God when He does not wish to sign'.

Finally, Giacomo Leopardi, in the 'Dialog of a peddler and a passer-by', writes:

> *Passer-by:* But what kind of life would you like, then?
> *Peddler:* Any kind, just as God would send it to me, with no other conditions.
> *Passer-by:* Any life at random, without knowing anything about it in advance, just as we don't know anything about the new year?
> *Peddler:* Precisely.
> *Passer-by:* That's what I would like too, if I were to live all over again; and that's what everyone would like. But this means that up until the end of this year, Fortune has treated everyone badly. And it's clear that everyone thinks that he was allotted more, and greater, evil than good; if given the chance to live the same life all over again, with all its good and all its evil, no one would want to be born anew. The life that's beautiful is not the life we know, but the life we don't know; not the past life, but the future. With the new year, Fortune will start treating you and me and all the others well, and the happy life will begin. Isn't it true?
> *Peddler:* Let's hope so.[1]

1 G. Leopardi, *Operette morali.* Here adapted from a translation for the initiative 'Randomness', by *The Italian Academy for Advanced Studies in America.*

2 It seems true, but some doubts arise

Randomness seems convincing, but at the same time we must admit a substantial order in all things. This applies to history also. Claude Lelouch's 'Bolero' is a film that shows us how a set of wholly unconnected elements – France, Russia, America, a dancer, a soldier, a musician, war and peace, and the course of time – relentlessly lead many people to Paris on the same day, to attend a ballet set to Ravel's famous 'Bolero'. Yet a small amount of disorder may be enough, as Poincaré explains[2], to change history. The flap of a butterfly's wing may cause a *catastrophe*[3], in the sense of the word as used by René Thom[4]. This last example, of the 'butterfly effect', is – in spite of the conviction of its inventor, of a good number of scientists, and of most of the general public – a phenomenon that could not be more deterministic[5].

In the macroscopic world, as in the micro-cosmos, individual events often seem to be random, while their aggregated whole seems foreseeable. Indeed, it is often computable, and consequently certain. Is it conceivable that individually stochastic events could somehow give rise to strictly deterministic laws, structures or organizations? We cannot say if and when a radioactive atom will disintegrate, but we can predict with astounding precision the number of atoms that will disintegrate in a gram of, say, plutonium, in a given unit of time. A termite mound is built without a project or a works manager, yet its structure is self-accomplished by the summation of the paths used to carry the materials with which it is built; ants collect food with a program that self-develops during the work; bees collect nectar with optimal runs that are the summation of the fastest random ones[6]. Many

2 H. Poincaré, *Science et Méthode*, Paris 1908.

3 E. Lorenz, *New York Academy of Science*, 1963.

4 R. Thom, *Stabilité structurelle et morphogénèse*, Interédition, Paris 1972.

5 G. Lanzavecchia, 'Riduzionismo e olismo, tra Galileo e Bateson', in *La Nuova Scienza*, Vol. 5. 'Scienza e tecnologia: al di là dello specchio', in U. Colombo and G. Lanzavecchia (eds.), *Libri Scheiwiller for Unicredit*, 2005, pp. 267-74.

6 R. Cervo & S. Turillazzi, 'Insetti interattivi' in *Galileo*, March 19, 2007, Year XI.

fractals in non-living systems[7] – such as coastal basins or star clusters[8] – are realized without an apparent code that gives instructions, as happens, instead, for plants. The sum of a set of – seemingly? – casual actions leads to well-organized, effective and regular structures and behaviour, which optimize resources, work, time, and the use of space.

What is to be said of serendipity[9], regarded as the symbolic example of chance, yet nowadays one of the most effective and certain tools for obtaining well-grounded results in all research fields[10], to the extent that it is considered *the* example – although certainly ambiguous – of the grimmest determinism? Serendipity is used to discover new drugs[11], and has represented the most important tool for the growth of the chemical industry[12]. It is even judged to be among the most valid ways to overcome the present Italian crisis[13]. And how would it be if nothing assimilable to chance were ever to occur? In such a case, we should be obliged to know the future – an idea that is abhorrent to most of us. Past, present and future would be presented simultaneously before us, without time's arrow, and perhaps also without entropy, and we should be obliged to live eternally.

7 B. B. Mandelbrot, 'Frattali', in *La Nuova Scienza*, Vol. 1, pp. 195-203, U. Colombo and G. Lanzavecchia (eds.), *Libri Scheiwiller for Credito Italiano*, Milano 2000.

8 'Is the universe a fractal?', in *The Scientist*, March 9, 2007.

9 A. Martinelli, 'Addio a Merton, teorizzò la "serendipity" e le conseguenze inattese', Corriere della Sera, February 25, 2003.

10 F. Bruno, 'Serendipity: fortunato incontro', in *Comunicazione Farmaco*, Year 2, Issue 1, 2004.

11 CHA Advanced Reports, 2015: Strategic Considerations for the U.S. Pharmaceutical Industry.

12 U. Colombo and G. Lanzavecchia, 'Luci e ombre della ricerca chimica in Italia' in G. J. Pizzorni (ed.), *L'industria chimica italiana nel novecento*, Franco Angeli, Milano 2006.

13 P. Bianucci, 'Crisi italiana – Diagnosi e cura', in *La Stampa – "Tutto scienze e tecnologia"*, May 11, 2005.

3 Does Laplace frighten us still?

Before the developments of modern physics – relativity and quantum mechanics, for instance – science considered the world precise and well determined. The mechanics of Galileo and Newton were busily explaining every motion (stars, terrestrial objects, particles, man-made engines, and so on), leading Pierre-Simon de Laplace to say:

> We may regard the present state of the universe as the effect of its past and the cause of its future. An intellect which, at a certain moment, knew all the forces that set nature in motion, and all the positions of all items of which nature is composed, if such intellect were also vast enough to submit these data to analysis, would embrace in a single formula the movements of the greatest bodies of the universe and those of the tiniest atom; for such an intellect, nothing would be uncertain, and the future, just like the past, would be present before its eyes[14].

Laplace is considered blasphemous even today, not so much because he places man very nearly at God's level, but because he wants to get rid of chance entirely. He does not wish to be arrogant, however. He uses the conditional of the verb, and points out the weak points of his reasoning. Nevertheless, with regard to the physics to which he is referring, he is right: the world is causally determined.

The growing awareness that each object interacts with all others to which it is somehow related (i.e., the whole universe), that each event depends upon all those with which it is interrelated (all that has happened since the big bang), and that each event, in turn, may cause all that which is to follow, has upset this framework and led to complexity and chaos[15]; we are dealing – unconsciously – with the most monstrous determinism, even if it is not describable by means of the calculus. But how does that differ from Laplace's view? Henri Poincaré was the first to examine this situation in a correct way, asserting that

14 P. S. de Laplace, Théorie analytiques des Probabilités, Paris 1812.

15 G. Lanzavecchia and M. Negrotti, In difesa della scienza. Patologie e promesse della razionalità nel XX secolo, Libri Scheiwiller, Milano 2002.

> A very small cause which escapes our attention, leads to a considerable effect that we cannot ignore, and then we say that it is due to chance[...]. Even if natural laws do not have secrets for us, we could know only approximately the initial conditions[...]. Forecast becomes impossible, and we have a casual phenomenon[16].

Approaches to, and theories of, complexity were often interpreted as the search, not for a hidden order – which was the purpose of classical science – but for a hidden disorder, in the wake of thermodynamics and entropy, Darwinian evolution, Maxwell's and Boltzmann's statistics, interpretation of time and its arrow, and hence of irreversibility.

Are complexity, chaos, disorder and inaccuracy a kind of non-Galilean science, or are they technology, engineering, application, indispensable for working out practical problems, even if only approximately? Is science still the search for simple, clear, 'eternal' laws that govern in a deterministic way the behaviour of matter, whatever it be – inanimate or living – wherever it be – on Earth or in the most remote reaches of the universe – whatever its state – nearly motionless with Newton's laws, or almost at light-speed with those of relativity – whatever its dimension – atom or subatomic particle, ruled by quantum mechanics, or enough matter to be subject to Newton's mechanics? In short, can one still think of science with the spirit of the Greeks[17], or is it mere application, without laws but only with rules that can be broken, contenting itself with models and useful theories, and consequently falsifiable?

4 In defence of science

Science is the only tool for the investigation and description of nature, matter, and the laws that govern them. It requires, above all, experimentation and the organized collecting of the results obtained, before fitting them into a coherent framework (using formulas, theories,

16 H. Poincaré, *Science et Méthode*, Paris 1908.

17 L. Russo, La rivoluzione dimenticata. Il pensiero scientifico greco e la scienza moderna, Feltrinelli, Milano 1997.

models), without introducing unverified or unverifiable hypotheses. In this way, it provides us with a tool that rigorously describes the observed phenomena, allowing us to make predictions. Science also requires the continual checking of experiments and theories, especially when new, more reliable results become available. The levels of observation must be unitary; the experiments simple – excluding every unnecessary 'accident' (according to Galilean terminology); the obtained representation reproducible; unnecessary hypotheses avoided, in accordance with Ockham's razor – so as to reach valid scientific and non-falsifiable results, at least within the reliability limits of the experimental data. This rigorous approach excludes complexity, multidisciplinarity, composition of knowledge, technology – that is to say, the 'falsifiable' parts of science, even if science does not repudiate or ignore them. Such aspects cannot be considered science, but rather approaches, attempts – at least as long as they remain inaccurate, indefinite, incomplete and falsifiable, even if indispensable for many activities. On the other hand, strictly Galilean and reductionist theories are still being created, which accurately describe nature, doing away with complexity and chaos, except when dealing with certain specific phenomena, such as the behaviour of the human heart, for instance, which is more or less physically or psychically stimulated. Such theories must not, above all, introduce preconceived hypotheses, concepts or reference points, except where observation – and not the requirements or suggestions of our own minds or cultures – demands it.

Amongst the new sciences we may recall, in particular, Albert Einstein's theories of relativity, both special and general, which describe motion and the nature of the universe (solar systems, space geometry, cosmology), and which, after a century of experiments, have survived intact, confirming their validity time and time again, however strange their predictions may be. Then there is Quantum Mechanics, which describes atomic and subatomic particles in a seemingly perfect way – so much so, that no other theory ever yet devised can verify and predict events with such precision. In this case, too, experimental verifications have always confirmed the theory. There are also the various theories of evolution, which frame all the transformation phenomena of inanimate and living matter. There is genetics, whose language provides the key with which to interpret all living structures. Finally, we may mention cosmology – still in a phase of elaboration – which describes the birth, evolution and future of the universe. All these theories were obliged to renounce prejudices: the nature of space and time; ideas of simultaneity, of causality, of the independence of parameters (position and motion, energy and time); the individuality of entities; construction of histories; and so on. Heisenberg wrote seventy pages to explain that one cannot speak of the 'orbit' or 'trajectory' of an electron – phenomena that are not observed with precision, and whose observation has no meaning.

Science is not without its faults, of course, some of which are due to a less-than-rigorous use of it, for example, as the explanation of an intrinsically complex world, known from time immemorial, when it is science's task to point out simplicity; or as the demonstration of 'intelligent design', when the experimental data describe evolution. Either we find experimental (and not just hypothetical) proof of intelligent design, or this last is not a part of science, but rather of philosophy or religion. One really critical fault of science is the inaccuracy of measurements, which form the basis of all experimentation, and the consequent need for a statistical approach, which leads to a probabilistic conception of phenomena, and hence of reality. A still worse shortcoming is the lack of 'perfect' macroscopic quantities: thus, for instance, if some particles are continuously being deposited upon, or evaporated from, a body, it cannot have mass, length, volume 'sub specie aeternitatis'. For elementary particles, instead, observation unavoidably affects that which is being observed. Finally, even at absolute zero matter has energy and motion.

5 Must one believe in intuition, reason or science?

Intuition and reason are unsatisfied by a science that does not possess absolutely certain tools – quantities and measures – and science responds by saying that those entities that people call certainties are not realizable, and therefore they do not exist. Luc de Clapiers, Marquis de Vauvenargues, writes that 'reason deceives us more often than nature', while John Locke, in *An Essay Concerning Human Understand-*

ing [Vol. IV, Ch. 14], affirms that 'God […] for the majority of that which interests us, has allowed us only the twilight […] of probability'. The author believes that reason is completely unreliable[18], as myriads of examples demonstrate, and also holds an impudent presumption: that of a mind that seeks to establish how the world is made up, and that wishes to teach – to nature, or, if you prefer, to God – what the laws of nature are, and also, in keeping with the current discussion, whether chance exists or not. However, these are arguments to which only rigorous observation – namely, that of science – can, perhaps, give an answer.

I believe that there may be an answer, notwithstanding our incomplete scientific knowledge of the world; but such an answer is probably, as Alan Cromer maintains[19], something less than we imagine. Moreover, it is unavoidably filtered by our minds – by reason – and therefore badly interpreted and interpretable, even when we believe we are behaving in the most rigorous and detached manner, without introducing prejudices, arbitrary hypotheses, or judgements on the merits of the observed facts and deductions drawn. An example is offered to me by an article written by my friend Massimo Negrotti[20], who cites an article by Edoardo Amaldi and Ettore Majorana, regarding the foundation of quantum mechanics, according to which (translated from Italian):

> Heisenberg had introduced this variable only as a formal artifice, and from a rigorously methodological point of view Majorana, at that time, was right in wanting to avoid the use of this new algorithm; but the subsequent history would show that, as sometimes happens, Heisenberg's choice, perhaps by chance, or perhaps by brilliant intuition, was the right one.

I have always thought that Schrödinger's procedure was non-scientific, as no very strong reasons existed to affirm that light and matter could be considered as waves and particles, while Werner Heisenberg adopted a rigorous procedure, considering only experi-

18 G. Lanzavecchia and M. Negrotti, In difesa della scienza. Patologie e promesse della razionalità nel XX secolo, Libri Scheiwiller, Milano 2002.

19 A. Cromer, *L'eresia della scienza*, Raffaello Cortina, 1996.

20 M. Negrotti, 'Chance would be a fine thing', elsewhere in this same volume.

mental data and their relations (emitted and absorbed radiation by atoms, for instance), without any introduction of hypotheses that were not experimentally well-founded. Indeed, Heisenberg could not otherwise have obtained the exact solution – not by chance nor by intuition – while Schrödinger arrived at his solution by way of an arbitrary intuition – and therefore, it seems, by chance.

6 What other tools does science have for facing the problem of chance?

Quantum physics upsets the traditional way of observing things. It begins by considering, not objects and states – whose existence and properties are prefigured – but actions and transformations, which instead are observed without being labelled in any way. In contrast to the traditional scientific method – as used by Schrödinger, for instance – this was the approach used by Heisenberg. It is an approach that may be, and already is, to some extent, applied to any science, be it physical or human. Is that enough to provide a convincing answer to the problem of chance, or must we resort to some other methodology? The 'imprecise' sciences – complexity, multidisciplinarity, technology – useful for giving specific answers to specific questions – are not useful for the consideration of general order relating to knowledge. General approaches are needed, such as those sought in physics to unify all the forces that govern the universe; in biology to frame, in a unique 'event', all life on our planet and in the universe; in sociology to extract rigid and coherent rules governing social organizations and groups. Physics has forces, biology has biochemical mechanisms, sociology has crowds, tribes and cities; but great overall visions are needed to explain – in the sense of providing coherent representations – the world in which we live.

Science has numerous approaches of a general nature, including: mathematics (an outcome of reductionism and perhaps the most extraordinary aspect of mechanism); global solutions, some of which have already been mentioned (DNA in biology, atoms in chemistry and physics), which can make the answers to our questions 'mechanical'; Ockham's razor (*'entia non sunt multiplicanda praeter necessitatem'*), along with its many variations (Principle of Economy, Laical Principle, Principle of Minimal Action, Principle of Frugality, Principle of Simplicity, Principle of Laziness, Principle of Minimal Effort, Principle of the Best)[21]; and finally, Fermat's principle, which brings with it a whole series of other principles, such as those of Lagrange, Hamilton and Feynman[22].

The 'Principle of Minimal Action', formulated by Pierre Louis Moreau de Maupertuis[23], is perhaps the most ingenious of these. The 'action' is a physical quantity defined as the product of kinetic energy and the time employed to cover a given trajectory, and the principle states that among all possible trajectories of a given duration, nature chooses that which minimizes the action – i.e., nature selects the minimum consumption of energy in a given time. The principle has always been verified without exception, and is one of the cornerstones of physics. The whole of mechanics can be derived from it, and it even plays a major role in Quantum Mechanics. Newtonian mechanics studies motion considered as infinitesimal jumps between points, balancing the forces at each one, and trying to translate all of this into a differential equation whose solution allows us to forecast the position of the point at every instant. The principle of minimal action, by contrast, can indicate the trajectory precisely, even if the initial conditions are not known, or if it is not possible to formulate a differential equation.

7 Does chance exist or not?

There are phenomena which, in the common way of thinking, are random, and they should be examined according to suitable criteria, in

21 'La grand unification de l'univers', (Internet 20/7/97, revised 20/3/02) <http://jac_leon.club.fr/gravitation/article-francais/f-34.html> <http://science-univers.qc.ca/cosmologie/ockham.htm>.

22 J. L. Basdevant, *Variational Principles in Physics*, Springer, 2007.

23 'Pierre Louis Moreau de Maupertuis, il principio della minima azione di Maupertuis', article from Wikipedia.

order to interpret them. So, for example, for radioactive atoms, the attribution of the thermodynamic time of the macroscopic world, or the astronomical time of planetary motions, would have no sense. These times have nothing to do with atoms. Unstable particles have an 'internal time' which is expressed, 'statistically', by the microscopic and macroscopic phenomena we observe. The indeterminacy of quantum mechanics is such for us, because we consider position and speed as independent variables, but it is not so for elementary particles. Indeed, measurement of their independent parameters gives exceptionally precise results. Many phenomena seem uncertain or undetermined individually, yet appear coherent and determined at a collective level, in a sort of teleological process that could depend on as-yet-unknown factors, or factors that we deny because they are inconsistent with our way of reasoning and with the world in which we are used to living.

In an article on defects and perfection[24], I pointed out the need for the existence of what are considered defects – that is, of the Gaussian curve taken as the symbol of reality by Negrotti[25] – and the inconsistency of perfection – wherein the entire Gaussian curve is squeezed onto a single value. I also showed how molecules, including genetic ones, can modify themselves by processes due to what Nobel Prize-winning biologist François Jacob has called molecular 'bricolage'[26], but also by improbable processes that increase, rather than reduce, their free energy. The mechanisms that modify molecules are due to encounters and collisions, thermal phenomena, cosmic radiation and a huge number of other factors – all well-established 'causes' that appear from one point of view as stochastic events, and from another as 'Darwinian evolution' and increasing complication – that is, as causal and teleological processes, if not 'intelligent design'.

If every event has a precise cause, then the world is a Laplacian one, and chance is a cultural construction. Or, there is a fundamental

24 G. Lanzavecchia, 'La perfezione non esiste o è un'anomalia', in *Nuova Civiltà delle Macchine*, XIX, No. 4, 2001.

25 G. Lanzavecchia and M. Negrotti, In difesa della scienza. Patologie e promesse della razionalità nel XX secolo, Libri Scheiwiller, Milano 2002.

26 F. Jakob, 'Les surprises du "bricolage moléculaire" ', in *Le Monde*, 4 janvier 2000.

indeterminacy, if not an incoherence, already at a sub-microscopic level, and an 'external' guidance – if we can call it that – of events in their coming together at a higher level, and the world is, at least in part, random.

8 Is a conclusion possible?

The author is inclined to believe in the non-existence of chance, in its most banal sense, for single events and for collective processes originating from them. Are free will and the freedom of evolution of the universe and of each man conceivable in a world without 'clinamen' (Democritus's random happenings)? As Gödel teaches us, every axiomatic set contains an infinity of undecidable propositions; and the universe, and we ourselves, are open, and not limited, systems[27], in spasmodic evolution[28], destined to alter every predictive process. Thus, there must continuously arise an infinity of new possibilities (and limits); that is to say, the world, being structured, open and evolving as it seems to be – and even if it is a Laplacian one – is potentially destined to every possible solution, implying an infinity of undecidable propositions and possibilities – and it is therefore free.

This is not really a conclusion, but rather a step towards a possible study to justify chance, keeping in mind that Gödel's reasoning may perhaps prevent any great unified solution.

Giuseppe Lanzavecchia, LCA – Laboratory for the Culture of the Artificial, Università di Urbino

27 G. O. Longo, *Da Golem a Goedel e ritorno*, Fucine Mute.

28 G. Lanzavecchia, 'Società: ieri, oggi e domani. Il mondo è cambiato: il passato è poca cosa se lo rapportiamo ai giorni nostri e al domani', in *Il Giornale dell'Ingegnere*, No. 6, April 1, 2007.

Stefano A. E. Leoni

Some questions on managing randomness by musical performer: about 'indeterminacy' (maybe: 'uncertainty') and 'cruces' (or knots, maybe: hubs?)

Abstract:

To make music, to compose it, but above all, to perform (and to listen to) it, is an activity strongly purpose to manage the randomness (the multi-randomnesses) that's implicit in the very structure of sound message. And music should represent an high level human activity of randomness management; whether that such a randomness should be caused by time *factor, or by the* tuning *one, or by the choice and by the organization of* intervals, *by the performing or interpretative options (particularly if dealing with old repertoire), or by the psychophysical interaction between the performer and his environment.*

One of the most intriguing areas concerning the connection among randomness, indeterminacy (uncertainty) and music is that of the interaction between two principles stated in a few years, at the beginning of the XX century. They're Heisenberg's Uncertainty Principle (1927) and Gödel's Incompleteness Theorem (1931).

But so far we are 'before' music, facing the sound. Music is something else, it's sound management, sounds' organisation, according to the most different criteria; music is a performing expressive structure, a sequential message streaming in the time, and having time as its dimension.

This is a kind of constructive negation of Randomness, of Chaos, differently positioned during the human history as history of artistic expressive structures and styles. In the West, from the Middle Ages to the XIX century, we witness a progressive tension to musical negentropy, proceeding at the same rate with the development of musical perceptive-receptive composing-organizing workings, so informing ones, aiming at the negation of the randomness, trying to teleologize it (from the compositions done using combinatory Mathematics, to the utilize of dices – Mozart, for istance – to the non-chaotic management of chaotic waves: stochastic composition or noise insertion). Nevertheless, it's actually important to explain connections and differences be-

tween randomness and uncertainty (indeterminacy), maybe starting from a prominent exemplum *such Cage is.*

Any way, we are 'deconstructing': music there is not, it is not (it's only in the instant, instant by instant between the chasm of the past and the shadow of the future); it isn't 'human' to play together, it isn't possible to manage in toto the variables that de-organize the performing pattern. Ergo: music is perfect illusion (by good fortune, probably).

In the end, here is another problem, another crux-hub-knot: the fear, the performance anxiety. In fact, nowadays the performance anxiety crux-hub-knot is one of the greater risk factor for the entropic disintegration of the performing message for academic musicians (not only for these ones, as we can read in the gossip column and in the crime news of the newspapers and magazines, about the stars *of pop music).*

Keywords:
music, indeterminacy, uncertainty, performer, time, tuning, performance anxiety, sequential (message), musical perception

Hermann von Helmholtz, the founder of the modern physiological acoustics studies, states in his Treatise (that 'was immediately greeted as a decisive contribute to knowing acoustical sensations and material foundations of music'[1]):

> When two musical tones are sounded at the same time, their united sound is generally disturbed by the beats of the upper partials, so that a greater or less part of the whole mass of sound is broken up into pulses of tone, and the joint effect is rough. This relation is called Dissonance. [...] But there are certain determinate ratios between pitch numbers, for which this rule suffers an exception, and either no beats at all are formed, or at least only such as have so little intensity that they produce no unpleasant disturbance of the united sound. These exceptional cases are called Consonances.[2]

1 A. Serravezza, *Musica e scienza nell'età del positivismo*, Il Mulino, Bologna 1996, p. 11.

2 H. v. Helmholtz, *Die Lehre von den Tonempfindungen als physiologische Grundlage für die Theorie der Musik, Braunschweig*, Vieweg 1863; Engl. Transl.: *On the Sensations of Tone as a Physiological Basis for the Theory of Music*, Fourth German edition, 1877; translated, revised, corrected with notes and additional appendix by A. J. Ellis. Reprint: Dover Publications Inc., New York 1954, p. 194.

Helmholtz, from his point of view, emphasize that Consonance is an abnormal occurrence, taking place under very special conditions, someone might say: random ones. That is because even if the fundamental tones have so different pitches not to produce audible beats, the higher harmonics can produce beats and get the whole resulting tone rough. So, for example, if two tones form a perfect fifth, the harmonic components in both the tones don't give any trouble; but if the relationship between the fundamentals is only approximately 2 to 3, those higher harmonics are not exactly alike, so produce beats and make the sound rough.[3]

To make music, to compose it, but above all, to performe (and to listen to) it, is an activity strongly purpose to manage the randomness (the multi-randomnesses) that's implicit in the very structure of sound message. And music should represent an high level human activity of randomness management; whether that such a randomness should be caused by *time* factor, or by the *tuning* one, or by the choice and by the organization of *intervals*, by the performing or interpretative options (particularly if dealing with old repertoire), or by the psychophysical interaction between the performer and his environment.

In the West this is verifiable during the whole history of this discipline, with one's own declension of ages, places, social roles, collective or individual typologies.

One of the most intriguing areas concerning the connection among randomness, indeterminacy (uncertainty) and music is that of the interaction between two principles stated in a few years, at the beginning of the XX century. They're Heisenberg's Uncertainty Principle (1927) and Gödel's Incompleteness Theorem (1931).

Roger Penrose, Professor of Mathematics at Oxford University and a physicist, now happily says, 'We cannot create any kind of new ar-

See also: B. Carazza and G. P. Guidetti, 'Helmholtz, la legge di Ohm e il problema dell'armonia', in *Giornale di Fisica*, XXX, 1989, pp. 207-214; G. Zanarini, 'Scienza e armonia. Hermann von Helmholtz e la spiegazione fisica della consonanza musicale', in *Nuova Civiltà delle Macchine*, XVI, 1-2, 1998, pp. 108-120.

3 H. v. Helmholtz, *Ueber die physiologischen Ursachen der musikalischen Harmonie*, 1857; Engl. transl.: H. v. Helmholtz, *Science and Culture: Popular and Philosophical Essays* (ed. by D.Cahan), University of Chicago Press, Chicago 1995, p. 72.

tistic sensitivity however we may accumulate many times of calculations. Art is a non-computable physics'. Someone – Iori Fujita – said: 'Music is a non-computable physics, too'. And then:

> over the centuries musicians, mathematicians, theorists, thinkers, experts and amateurs have been suffered from the comma, which is the difference between a perfectly tuned octave and the octave resulting from a tuned circle of fifths. Many great people have been trying to create the perfect scale in vain. Mathematics easily proves that perfection is not possible. Any solution does not exist. Musicians, especially pianists, have been accused of using the Equal Temperament for their pianos because the Equal Temperament is said to be an anti-musical compromise which leaves each key equally damaged and none perfectly in tune. This comma has put a curse on music.[4]

The *comma* is the irrational element, not allowing a relationships between the octave and a set of fifth (but also a set of thirds, or any other 'perfect' interval). Here is its maths reason for this aporia:

$$\left(\frac{3}{2}\right)^{n} \neq \left(\frac{2}{1}\right)^{m} \text{ or, also } : \left(\frac{5}{4}\right)^{n} \neq \left(\frac{2}{1}\right)^{m} \text{ etc}$$

In spite of this, the history of Music Theory (and Practice) is marked by constant attempts, sometimes shrewd, sometimes naïve, sometimes abstruse ones, to find a solution for this problem. And in spite of this all, music has gone on, all over the place, with some 'immeasurable' intervals regarded as preferential (of course, Helmholtz said that it's a 'natural', physiological attitude, even if – later on – a cultural one)[5].

It's something like a real philosophical 'unknownness' theorem (H. Putnam), sprung from Einstein theory of Relativity and from Gödel and Heisenberg ideas which, from the spheres of phisics and mathe-

4 I. Fujita, 'Uncertainty Principle for Temperament' at: <http://www.geocities.jp/imyfujita/wtcuncertain.html>.

5 'Il fatto che una combinazione [di suoni] sia più aspra o più dolce di un'altra dipende unicamente dalla struttura anatomica dell'orecchio, e non ha nulla a che fare con ragioni psicologiche. Ma il grado di asprezza che un ascoltatore è disposto ad accettare come mezzo di espressione musicale dipende dal gusto e dall'abitudine: e infatti il confine tra consonanza e dissonanza è stato spesso modificato', H. v. Helmholtz, *Die Lehre*, cit. p. 234.

matical logic, have been apply in knowledge fields so far from their own. So it's no accident that a few phisicists-musicians have thought of connecting Heisenberg's Uncertainty Principle and some aspects of music. Among these the Australian Joe Wolfe and the Japanese Iori Fujita.[6]

Joe Wolfe, for example, states:

> Your results might depend upon your hearing, how much background noise there is, and how hard you concentrate. But you probably found that, when the frequencies differed by 3 Hz, you needed about a third of a second. When they differed by 1 Hz, you needed about one second. So, roughly speaking, if the frequencies differ by Δf, then you need a time of 1/Δf to notice. In other words:
>
> $\Delta f.\ \Delta t > \sim 1$ or, in non-mathematical language:
>
> (time taken to measure f) times (error in f) is about one or greater. I call this:
>
> The musician's uncertainty principle. Because musicians know this, qualitatively at least. If the chord is short, or if you are playing a percussive instrument, the tuning is less critical. In a long sustained chord, you have to get the tuning accurate. And of course oboists in orchestras play notes for tens of seconds while the other instruments tune carefully before a concert.
>
> Heisenberg's uncertainty principle. Now in quantum mechanics and atomic physics, the energies of a photons is hf, where h is Planck's constant. So a measurement of the energy corresponds to a measurement of the frequency, and that, as we have seen, takes time. Mutliplying our previous inequality by h on both sides gives us
>
> $$\Delta\,(hf).\ \Delta t > \sim h$$
>
> (uncertainty in energy) times (uncertainty in time) is greater than about h.
>
> So far this calculation is just an order-of magnitude. One can imagine doing a bit better than $\Delta f.\ \Delta t > 1$ by careful measurement. (Have a look at the diagrams on What are interference beats?).
>
> Heisenberg's uncertainty principle for energy is usually written with an extra factor of 2p.
>
> $$\Delta E.\ \Delta t > \sim h/2p.$$

6 S. A. E. Leoni, 'Musica, meccanicismo e tempo', in S. A. E. Leoni e P. A. Rossi, *Manuale di Acustica e di Teoria del suono*, Rugginenti, Milano 2005(2), pp. 207 ss.

Heisenberg's uncertainty principle for momentum is analogous. Let's consider the spatial frequency F (which is defined as the number of cycles per unit distance) rather than temporal frequency (number of cycles per unit time). F is just the reciprocal of the wavelength, l. The same argument about beats in this case gives (for motion in the x direction)

$$\Delta F.\ \Delta x = \Delta\ (1/l).\ \Delta x >\sim 1$$

The momentum of a photon (or anything else) is p = h/l so, multiplying the above equation by h gives

$$\Delta p.\ \Delta x >\sim 1$$

which is usually written as

$$\Delta px.\ \Delta x >\sim h/2p.$$

(uncertainty in momentum) times (uncertainty in position) is greater than h/2p.

Practical consequences of the uncertainty principle.

h is very small ($6.63\bullet 10^{-34}$ Js), so the consequences of the uncertainty principle are usually only important for photons, fundamental particles and phonons. (See, for example, this example using a cricket ball.) There are, however, many physical processes whose evolution with time depends sensitively on the initial conditions. (Sensitivity to intial conditions is fashionably called chaos.) The uncertainty principle prohibits exact knowledge of initial conditions, and therefore repeated performances of such processes will diverge. (Physicists will also tell you that one cannot have exact knowledge anyway, for a variety of practical reasons, including the fact that you don't have enough memory to record the infinite number of significant figures required to record an exact measurement.)

Philosophical consequences of the uncertainty principle. Some philosophers regard the consequences of the uncertainty principle as having a more fundamental importance. The argument goes like this: if one could know exactly the position, velocity and other details, one could, in principle, compute the complete future of the universe. Since one cannot know the position and momentum of even one particle with complete precision, this calculation is impossible, even in principle. Most scientists find this a trivial argument. A memory capable of storing all this information would be as complex as the universe, and then the contents of that memory would have to be included in the calculation, and that would make the amount of information greater, and that information would have to be stored..... We rather point out that all of that

> information is actually contained in the universe which, as an analogue computer, is computing its own future already.[7]

More or less the same that Iori Fujita states; he extends these evaluations to the Fourier Transform Theory too.

> Can this [the Heisenberg Uncertainty Principle] be applicable to the sound wave? This is for the behavior of single subatomic particles, isn't it? The Planck's constant 'h' is too small to be took into consideration in our daily life.
>
> $$h = 6.6261 \times 10^{-34}\ \text{Js}$$
>
> A tennis ball (about 50g) with a 180km/h (50m/s) speed which was served by Ms. Martina Hingis would have any uncertainty about its location? Actually No! In case of $\Delta v=0.01$m/s, the quantum deviation Δx is about $2*10^{-31}$m or 0.0000000000000000000000000002 mm.
>
> Sound is the result of periodic fluctuations in air pressure propagating through the ordinary atmosphere. The sound waves are in the ordinary world. You may think that the Uncertainty Principle has nothing to do with music. But the formula can be transformed into the relation between time and frequency of waves.
>
> $$\Delta x \Delta p \geq h/(4\pi)$$
>
> $p = mv \rightarrow \Delta p = m\,\Delta v$ $\quad$ v = velocity (m/s)
> $E = m\,v^2/2 \rightarrow \Delta E = m\,v\,\Delta v = m\,\Delta v\,v$
> $\Delta E = \Delta p\,v = \Delta p\,(\Delta x/\Delta t) = (\Delta p\,\Delta x)/\,\Delta t$
> $\Delta E \geq (h\,/(4\pi))/\,\Delta t$
> $\Delta E\,\Delta t \geq h\,/(4\pi)$
> $E = h\,\nu \rightarrow \Delta E = h\,\Delta\,\nu$ $\quad$ ν = frequency (1/s) = *f*, E for a photon
> $h\,\Delta\nu\,\Delta t \geq h\,/(4\pi)$
>
> $\Delta\nu\,\Delta\,t \geq 1\,/(4\pi)$ Here, there is no 'h'. $\quad$ $\Delta\nu \geq 1\,/(4\pi\,\Delta t)$
> $\Delta t \geq 1\,/(4\pi\,\Delta\nu)$
> Δt and $\Delta\nu$ are the standard deviation of t and ν.
>
> The position and momentum of a wave in space have been changed in form to the time and frequency of a wave. Here is a sound wave as a time-varying signal. Now we have found out the fact that it is impossible for us to know the

7 J. Wolfe, 'Heisenberg's uncertainty principle and the musician's uncertainty principle', at: <http://www.phys.unsw.edu.au/jw/uncertainty.html>.

exact frequencies in our sound wave at an exact moment in time. When $\Delta\nu$ is 0.1Hz, Δt should be more than $1/(4\pi \cdot 0.1)$ sec which is about 0.80 sec.

Middle C of the equal temperament is calculated approximately as 261.6256Hz. If you want to determine C with this precision 0.0001Hz, you need Δt of 800 sec, or 13 min 20 sec. If you want to get the exact Middle C, you need an infinite time and a continuous wave of C.

In other words, theoretically we cannot get the perfect fifth tone with a frequency f (that is ν) from a root tone with a frequency f_0 by calculating $f = f_0 \times 3/2$.

In another treatment, the 'uncertainty' of a variable is taken to be the smallest width of a range which contains 50% of the values, which, also in the case of normally distributed variables, leads to a lower bound of

$\Delta\nu\ \Delta t \geq 1/(2\pi)$ for the product of the uncertainties. For different types of wave packets and for other treatments, the uncertainty can be set to a much lower bound.

$\Delta\nu\ \Delta t \geq 1/(4\pi)$ is for the case of the normally distributed variables or the minimum wave packet like a photon and a wave with the standard deviation form.

$\Delta\nu\ \Delta t \geq 1$ is the ordinary interpretation for the frequency and time.

In this physical world there can exist no temperament based on the ratios of whole numbers. Even harmony must live with this 'uncertainty'.

The Fourier Transform Theory:

You may say again that it is only for the subatomic world. But the Fourier Transform Theory reveals that the sound wave cannot get rid of the Uncertainty of Time and Frequency.

Fourier Analysis for music is based on the concept that sound waves can be approximated by a sum of sinusoids. The frequency of each sinusoid in the series is an integer multiple of the frequency of the basic signal. It is said to be the same as the harmonics of the original waveform.

$$f(t) = \tfrac{1}{2}a_0 + \sum_{n=1}^{\infty}\left[a_n \cos(n\omega t) + b_n \sin(n\omega t)\right]$$

It is a very useful theory for us to study timbres. There are sawtooth, square, trangle, sine and other types of wave form. Each form corresponds with a certain timbre. By arranging an and bn, we can create any type of wave form.

The equation below shows how a square wave can be made up by adding together pure sine waves at the harmonics of the fundamental frequency.

$$f(t) = \sin(\omega t) - (1/3)\sin(3\ \omega t) + (1/5)\sin(5\ \omega t) - (1/7)\sin(7\ \omega t) + ...$$

In general; $a_o = 0$ and $b_n = 0$

$$f(t) = a_1 \sin(\omega t) + a_2 \sin(2\ \omega t) + a_3 \sin(3\ \omega t) + a_4 \sin(4\ \omega t) + ...$$

This $f(t)$ is defined within a certain area. And It should be periodic. So, $f(t) \neq 0$ outside of the area. The sound can't stop while accumulating sin($n\omega t$)s.

This theory is for a periodic and continuous wave. Sound waves are approximately periodic within the time-length while they exist and to some extent continuous before they fade away. The Fourier Analysis is setting forth 'the infinite continuity of the wave' as a premise. At least it is a theory for a stationary wave. Sounds are to be classified as non-continuous and pseudo-periodic waves. In other words, it is difficult to describe an actual sound of a limited duration by integer multiples of the frequency of its basic signal.

Then comes the Fourier Transform Theory's turn;
A physical signal, such as sound pressure can be represented as a continuous function of time. This is the time domain representation of the sound. There is an equally valid frequency domain representation.

$f(t)$ = a sound wave; for example; the simplest one:

$$f(t) = A(t)\sin(2\pi\, v\, t);\ v = \text{frequency}(1/s)$$

$$F(\omega) = \int_{-\infty}^{\infty} f(t)\, e^{-j\omega t}\, dt$$

$$\omega \leftrightarrow t \quad F(t) \leftrightarrow 2\pi\, f(-\omega)$$

$$f(t) = \frac{1}{2\pi}\int_{-\infty}^{\infty} F(\omega)\, e^{j\omega t}\, d\omega$$

$$\omega = 2\,\pi\, v$$

$F(\omega)$ equals the integral from minus infinity to infinity of *f* of *t* times the exponential to the minus $j\ \omega\ t\ dt$.
The Fourier transform separates a waveform into sinusoids or sine curve functions of different frequency which sum to the original waveform. It is typically thought of as decomposing a signal into its component frequencies and their amplitudes. The magnitude of this function $F(\omega)$ is normally called the 'frequency response'. And the magnitude of the resulting complex-valued function $F(\omega)$ represents the amplitudes of the respective frequencies ω and shows the frequency distribution.[8]

Fortunately there is a difference between 'physical' tuning and 'perceived' tuning; a reflection of Weber-Fechner law, which states that the ration between physical entity and sensation isn't linear, and that the intensity of a physiological sensation is proportional to the stimu-

8 I. Fujita, *cit.*

lus' logarithm. Not only: there is a set of acoustical-perceptive 'servomechanisms' that goes beyond that law: for example the *Mel* curve.

So far we are 'before' music, facing the sound. Music is something else, it's sound management, sounds' organisation, according to the most different criteria; music is a performing expressive structure, a sequential message streaming in the time, and having time as its dimension. Going by what H. H. Eggebrecht says:

> One of the forms of course is the time of course oriented to an aim, teleological. It has an outstanding rôle in Western music. The key of a Medieval tune is given by the *finalis*; a dissonance tends to resolution; a *clausula* or cadence moves towards the tonic closing compound; an opera or a symphony gravitates to the *Finale* [...] These teleological-oriented musical structures correspond to the rule of ideas, canons and plans generally oriented to an aim.[9]

This is a kind of constructive negation of Randomness, of Chaos, differently positioned during the human history as history of artistic expressive structures and styles. In the West, from the Middle Ages to the XIX century, we witness a progressive tension to musical negentropy, proceeding at the same rate with the development of musical perceptive-receptive composing-organizing workings, so informing ones, aiming at the negation of the randomness, trying to teleologize it (from the compositions done using combinatory Mathematics, to the utilize of dices – Mozart, for instance – to the non-chaotic management of chaotic waves: stochastic composition or noise insertion).

It's actually important to explain connections and differences between randomness and uncertainty (indeterminacy), maybe starting from a prominent *exemplum* such Cage is.

In John Cage indeterminacy-uncertainty is a 'constructive', even a 'meaning-producing' element, involving various parameters, from the time one (suspension of time vector) to the tone-colour one (prepared piano, for instance), to the *aleatoric* composition.

As to the suspension of the 'time' relations, Cage's considerations are cardinal in the XX century musical thought. Not only his own

9 H. H. Eggebrecht, 'Tre pezzi brevi. Musica come tempo', in *Il Saggiatore musicale*, VII, 2000, p. 390 <http://www.muspe.unibo.it/period/saggmus/attivita/doc/musica.htm>.

ideas, those he took from Satie and oriental philosophies, but also those of his circle, such as Morton Feldman's. This one, in his essay 'Between categories', keep his distance from the idea of composition as a construction of sound connections in the time and states: 'I prefer to think of my work as: between categories. Between time and space. Between painting and music. Between the music's construction, and its surface.'

Morton Feldman's music 'is' sound, the becoming-memory of sound, becoming-memory of time, becoming-memory of silence. Becoming-memory in the space of the slowly striding time. Bordering on the essential nature of the medium, sound in time, between fleeing sound patterns constantly changing place and breathing silences, Feldman's music draws the image of time, 'time in its unstructured existence'. Generative – not figurative. Feldmans unique way of working hides 'a conscious attempt at "formalizing" a disorientation of memory', where escaping from a dialectic subjected composing, the 'unification of opposites' is the aim.

By writing music in which the memory operates as the generative force of oblivion itself, Feldman escapes from narrative construction, music as metaphor for history, structure and memory. Feldman follows Cézannes credo: 'This was not how to make an object, not how this object exists by way of Time, in Time or about Time, but how this object exists as Time. Time regained. Time as an Image.'[10]

Up to Milton Babbitt (in part), to Conlon Nancarrow and then to *Fluxus* and to the Minimalists, those started from Cage's experience, developping it autonomously, leaving out indeterminacy-uncertainty matters, and focused time considerations. Up to Robert Ashley (1930).

The problem of indeterminacy is a key point in a poetics 'that identifies in the exploration of the sound phenomenon as a metaphor of life multiplicity, and associates a general and essential cognitive apti-

10 At: <http://212.123.5.64/www.orpheusinstituut.be/default.asp>. See: B. H. Friedman (ed.), *Give My Regards to Eighth Street: Collected Writings of Morton Feldman*, Exact Change, Cambridge 2000.

tude with the inclination for crossing'.[11] Cage composing doctrine is based on 'moving towards multiplicity', on 'moving away from unity', fixing a conceptual link between sound indeterminacy and random constructive plans: composer's subjectivity withdraws, favouring life as expression of both the 'indeterminate' and 'indeterminable' multiplicity: 'I think it's wonderful that life breaks us off.'[12]
As someone wrote:

> Indeterminacy refers to aesthetic values connected with perception, and it's used as an *expressive* medium for obtaining a particular result. Indeterminacy rules out exact determining of some parameters (pitch, duration, timbre, loudness) to give a margin of freedom to the performer; this freedom becomes not only interpretation, but gets to be part of the very formal identity of the work. So indeterminacy is just the opposite – but also the 'complement' – of randomness, that instead adopts a method becoming the necessity. The randomness relies on *method logic*, the indeterminacy relies on *performer sensitivity*.[13]

When the author withdraws himself, also *opus* withdraws: as a finished artistic object, as an *unicum* eternally and necessarily, – so to speak: 'ethically' – replicable, highlighting the aesthetic nature of creative phenomenon in its 'being-for-the-senses', in its getting on a level with the perceiver and of his 'multiplicities'. Suond shows itself in a discontinuous way in a pre-artistic space that's the *silence*: so the radical results, that lead Cage to dialectically confront with silence (see *4'33"*, in 1952), underline a value, nihilist crisis: silence is *indeterminacy that receives all the not yet expressed possibilities*; it's the place where sound phenomenon arises as a vital phenomenon.

With these arguments the problem of the ticklish, key connection 'composer-work-performer-(public)' comes out; about this matter, we have got to remind – as for random element *management* – a few main-themes to think over.

As Eric Clarke states:

11 F. Aste, 'Il materiale e il processo compositivo tra indeterminazione e necessità. Le *Sonatas and Interludes* per pianoforte preparato di John Cage', at: <http://users.unimi.it/~gpiana/dm9/aste/articolocage.pdf>.

12 J. Cage, *Lettera a uno sconosciuto*, Socrates, Roma 1996, p.50.

13 F. Aste, *cit.*

> Musical performance at its highest level demands a remarkable combination of physical and mental skills. It is not uncommon for concert pianists to play at speeds of ten or more notes per second in both hands simultaneously, in complex and constantly changing spatial patterns on the keyboard, and with distinct patterns of rhythm, dynamics and articulation. Equally, a performer has to have an awareness and understanding of the immediate and larger-scale structure of the music itself, an expressive 'strategy' with which to bring the music to life, and the resilience to withstand the physical demands and psychological stresses of public performance.[14]

The job of the performer, of the *musical performer* (not the mere *executor* Stravinskij would like) is the formulating of interpretations, on the basis of a series of instructions, considering the difficult and faint relation, the 'slippery', sometimes 'chaotic' or 'random' relation, between instructions, formulas, and the wish of exerting his/her own interpretative intuitions.

> Musical interpretations are communicated through the expressive parameters of timing, dynamics, articulation and timbre, among others. A score might contain a variety of expressive indications to aid the performer's interpretative choices, but, ... expressive notation lacks precision[15]

because presumed, during the history, a contiguity between composer and performer such as yield co-authorial codes partly passed through a crisis in the XIX and in the XX century.

> Performers – as Stefan Reid underlines – must tread the difficult path between the need to respect the score, which represents the composer's intentions, and the desire to exercise their own creative insights.

Of course with all the necessary fine distinctions, with the HIP and the criticism towards *authenticity.*

> In addition, aesthetic ideals vary from person to person and according to the fashions of the day. Interpreting music is therefore a highly subjective process and resistant to prescriptive recommendations. Whereas the advice offered in

14 E. Clarke, 'Understanding the psychology of performance', in J. Rink (ed.), *Musical Performance. A Guide to Understanding*, Cambridge University Press, Cambridge 2002, pp. 59-72.

15 S. Reid, 'Preparing for performance', in: J. Rink, *cit.*, p. 106.

> the pedagogical and psychological literature contains specific instructions for the development of technical expertise, interpretative advice is, not surprisingly, often more varied and less detailed.[16]

Interpreting music is a process on the border of indeterminacy, incompleteness (in the Gödel sense of the word) and randomness.

> The starting point for performances of Western music is normally a score, which consists of a series of instructions of varying degrees of indeterminacy that the performer must then translate into sound. The *indeterminacy* inherent in Western musical notation means that the decoding of the score requires considerable interpretative input and insight from the performer. Consequently, no two performances of a work will be the same.[17]

The more decisive factor for the difficulty of management of random elements in music is time as ineluctable vector for realizing the musical project in sound. If the score is a pre-musical object (and project), the performing (and listening) act is 'sequential' and lives only in 'being-for-the-moment-reminding-the-past-foretelling-the-future'; Michael Tree, a member of the Guarneri String Quartett, says: 'every moment of our playing is conditioned by what has just occurred or by what we think is about to occur'.[18]

> Musical interaction can be planned to a certain extent in rehearsal: performers might wish to work out who will follow whom during a particular passage, or who will take the lead in the ensuing passage. This type of conscious planning can usefully assist coordination, but it does not account for the moment-to-moment hunting and cooperating that necessarily go on throughout performance.[19]

The anticipation of each beat and the reaction to its production are defined in effect by the nature of the musical interaction between the performers and therefore are very fluctuating. In the ensemble per-

16 *Idem*, pp. 106-07.

17 *Idem,* pp. 102.

18 E. Goodman, 'Ensemble performance', in J. Rink, *cit.*, p. 153; quotation from D. Blum, *The Art of Quartet Playing: The Guarneri Quartet in conversation with David Blum*, Cornell University Press, Ithaca 1986, p. 20.

19 *Idem*, p. 155.

formance too, if the definition of a *tactus*, of a main beat (i.e. the overall *tempo*) provides a benchmark for keeping time, and 'functions as the ensemble's clock, for it provides a source of coordination and controls the beat ticking inside each musician' producing

> a 'shared common timekeeper' in the overall tempo of the music ... While the main beat is an important yardstick for keeping time, local considerations might inspire individual musicians in an ensemble to subdivide or lengthen the beat in different places. Here, the bar acts as an important structure for coordination, since it is a unit of time around which the main pulse can be individually organised. Performing in time with other people requires more than just the ability to count and to realise the overall tempo, for the execution of each beat in performance needs to be carefully administered. This involves two main skills: anticipation and reaction. Ensemble performers carry out complex predictions that are intimately bound to reactions gained through feedback[20]

and through a coordinated repetition of the musical 'athletic act', in rehearsal.

> There are several factors to be taken into account when trying to achieve coordination in a musical ensemble, not least the physical differences between instruments (including the human voice). Musicians need to be aware of the amount of time it takes for a note to 'speak' on a particular instrument.[21]

Not to mention of the space placing of the performers. So, art of performing is the art of creating the illusion of synchrony, the illusion of a perfect ensemble:

> Successful timekeeping ideally results in the coordination of notes between individual musicians in an ensemble, but the reality of performing together is not so simple. In fact, the execution of notes at exactly the same time by a group of musicians is beyond the limits of human skill and perception: there will always be minute discrepancies in timing – that is, asynchronisation – between the notes intended to be performed simultaneously.[22]

20 *Idem*, p. 154.
21 *Idem*, p. 155.
22 *Ibidem.*

We are 'deconstructing': music *there is* not, it *is* not (it's only in the instant, instant by instant between the chasm of the past and the shadow of the future); it isn't 'human' to play together, it isn't possible to manage *in toto* the variables that de-organize the performing pattern. *Ergo*: music is *perfect* illusion (by good fortune, probably).

In the end, here is another problem, another crux-hub-knot: the fear, the performance anxiety. Recent studies, well summarized in an essay by Elisabeth Valentine, inform us of the increasing difficulties among professional musicians and of the growing use of physical-pharmacological treatments directed to managing anxiety by the performer.[23]

What's performance anxiety?

> Performance anxiety, commonly known as 'stage fright', is an age-old problem, but interest in its nature, causes and cures has intensified over the last fifteen years with the burgeoning of clinics, conferences and journals devoted to performing arts medicine. Music performance anxiety has been defined as 'the experience of persisting, distressful apprehension about[,] and/or actual impairment of, performance skills in a public context, to a degree unwarranted given the individual's musical aptitude, training, and level of preparation'. A comprehensive account of performance anxiety thus needs to include physiological factors (such as heart rate and blood pressure), behavioural measures (of anxiety and the quality of performance) and self-reports (of thoughts and feelings). The reactions of these three systems may not be correlated. In particular, it is common to have physiological symptoms without either of the others. Correlation is more likely to occur in states of high anxiety.[24]

Beside the larger and larger use of beta-blockers (beta-adrenoceptor blocking agents), alcohol and sedatives, some techniques are carried: combine physical relaxation with mental alertness, autogenic training, behaviour therapies, *cognitive systematic desensitisation, biofeedback*; alexander technique, the *Inner Game*, or more psychological-psychotherapeutic techniques, such as the *cognitive behaviour therapy* or the *timing of anxiety*; and more, treatment packages, combining different techniques.

23 E. Valentine, 'The fear of performance', in J. Rink *cit.*, pp. 168-182.

24 *Idem*, p. 168.

Anxiety is a terrific de-organizing, disintegrating element heavy-inserting random and disinformation in a performance.

> The symptoms of performance anxiety are well known and are of three kinds: physiological, behavioural and mental. The physiological symptoms of increased heart rate, palpitations, shortness of breath, hyperventilation, dry mouth, sweating, nausea, diarrhoea and dizziness are the result of over-arousal of the autonomic nervous system. This flight-fight response, which assisted our hunter-gatherer forebears in fleeing large animals, is highly detrimental to musicians requiring dexterity and fine muscular control over their instruments. Trembling limbs and slippery fingers are likely to hinder rather than help the performer. In addition, this autonomic arousal may have become associated with fear as a result of past experience. Increased arousal generally leads to a narrowing of the focus of attention, which may also be deleterious. The behavioural symptoms of performance anxiety may take the form either of signs of anxiety, such as shaking, trembling, stiffness and dead-pan expression, or of impairment of the performance itself. The mental symptoms are subjective feelings of anxiety and negative thoughts about performing. Rather than fear of performance per se, it is fear of public performance that is at issue, with the risk of negative evaluation and consequent loss of self-esteem.[25]

In fact, nowadays the performance anxiety crux-hub-knot is one of the greater risk factor for the entropic disintegration of the performing message for academic musicians (not only for these ones, as we can read in the gossip column and in the crime news of the newspapers and magazines, about the *stars* of pop music).

> It is important to distinguish between beneficial and detrimental kinds of anxiety – or to be more precise, between reactive, maladaptive and adaptive anxiety. Reactive anxiety, the result of inadequate preparation, is realistic and is best dealt with by music analysis and rehearsal. Anxiety is widely regarded as deleterious, but every performer knows that a certain amount of arousal is beneficial to performance. Performance is generally best at moderate levels of arousal: if arousal is too low, the performance will be dull and lifeless; if it is too high, the performer and the performance may come to pieces. This can be represented in the shape of an inverted U (plotting quality of performance as a function of arousal) and is known as the Yerkes-Dodson law.[26]

25 *Ibidem.*
26 *Idem*, p. 170.

Until the anxio-genetic stimulus is low, the relation between *arousal* and *performance* follows the Yerkes-Dodson curve, but when it is high, it follows the catastrophe model: when *arousal* increases, the *performance* is victim 'of a catastrophic decline, from which it is difficult to recover'.[27]

This is performing chaos, disorder, that produces perceptive chaos and disorder: interpretative mental models broken up by the impossibility of managing one's mind and one's body tuned in to environmental reality. A failed *performance*, then.

'If Orpheus did tame wild beasts with his lyre – and who are we to say otherwise? – he must have known exactly what he was doing as a performer'.[28]

But shamanic magic operates on the *naturalia*, zeroing the indeterminacy, the uncertainty, or the randomness in music!

Bibliography

J. Dunsby, *Performing Music: Shared Concerns*, Clarendon Press, Oxford 1995.

B. Green and T. C. Gallwey, *The Inner Game of Music*, Doubleday, Garden City, N.Y. 1986.

I. Fujita, 'Uncertainty Principle for Temperament' at <http://www.geocities.jp/imyfujita/wtcuncertain.html>

K. Havas and S. Fright, *Its Causes and Cures, with Special Reference to Violin Playing*, Bosworth, Londra 1973.

S. A. E. Leoni and P. A. Rossi, *Manuale di Acustica e di Teoria del suono*, Rugginenti, Milano 2005(2).

D. Meichenbaum, *Stress Inoculation Training*, Pergamon, New York 1985.

J. Rink (ed.), *Musical Performance. A Guide to Understanding*, Cambridge University Press, Cambridge 2002.

E. Ristad, *A Soprano on Her Head: Right-side Up Reflections on Life and Other Performances*, Real People Press, Moan, Ut. 1982.

27 *Ibidem.*

28 J. Dunsby 'Performers or performance', in: J. Rink, *cit.* pp. 225-235.

C. E. Seashore, *Psychology of Music*, Dover, New York 1967(2), (McGraw-Hill, 1938).

J. A. Sloboda (ed.), *Generative Processes in Music: The Psychology of Performance, Improvisation, and Composition*, Clarendon Press, Oxford 1988.

J. A. Sloboda, *The Musical Mind. The Cognitive Psychology of Music*, Oxford University Press, Oxford 1985, It. transl, *La mente musicale*, Il Mulino, Bologna 1988.

R. Triplett, *Stage Fright: Letting it Work for You*, Nelson-Hall, Chicago 1983.

G. D. Wilson, *Psychology for Performing Artists: Butterflies and Bouquets*, Jessica Kingsley, London 1994.

J. Wolfe, 'Heisenberg's uncertainty principle and the musician's uncertainty principle', at http://www.phys.unsw.edu.au/jw/uncertainty.html.

Stefano A. E. Leoni, Conservatory of Music of Torino and IMES-LCA, Università di Urbino

Giorgio Mainini

Randomness: a course in survival

Abstract:

This article illustrates, with some considerations and a few examples, that the need is acute for a better understanding of probability and statistics, in order not to be fooled by unknowledgeable and biased description of events. It further attempts to demonstrate that the places to introduce a feeling for probability and statistics are the early, compulsory schools, such as primary and junior high. Lastly, it claims that this introduction should be not only the responsibility of teachers, whose fundamental and continuing education has been traditionally neglected in regard to the development of that feeling, but also of their educators.

Keywords:
statistics, statistical events, probability, evaluation of probability, continuing education of teachers.

1 Introduction

Lottery, screening for breast cancer, death rate, DNA analysis, bus stops, what do all these have in common? If one plays *tern*, what is the probability of guessing the numbers? And if guessed, how much will that pay? If a woman has screening to detect breast cancer and the test is positive, what is the probability that she really has cancer? If the death rate for a seventy-year-old person is 2.5%, what is the probability that a certain seventy-year-old person will give his or her soul before the end of the year?

If, after a test to determine paternity, a man is recognized as the father of a child, what is the probability that he truly is the father? If a bus has passed by a given stop, what is the probability one has to wait more than five minutes for the next one to come by?

In everyday life, a large amount of phenomena occur with a certain *probability*. We should, therefore, be able to evaluate it, if not with accurate precision – which is, in most cases, impossible – at least within reasonable bounds. However, field research has shown that the subjective perception for the probability of a given event is strongly distorted. And this is a serious shortcoming, since several exploiters profit from this ignorance, institutional ones such a lotteries as well as quack sellers of certainties.

Let us see just one example close to home. Next to all casino roulettes, there is a bright column with the indication of all recent hits. However, a moment of thought should tell that a ball does not have any memory! Thus, the probability that, at the next throw, the ball hits 23 – which came out 3 rounds ago – is exactly the same as hitting 18, which has not come out for 30 rounds. But the bright column suggests otherwise! Hence, many families have been ruined, induced to agonizingly bet on a rare number at the state lottery, because of repeated articles about this rarity in supposedly serious newspapers and cabalistic magazines.

2 Some Answers

This is not a treaty in probability calculus in order to attempt to answer the questions of above. The ambition of this writing is to claim the necessity for the introduction of a feeling for probability at the level of primary and junior high schools, which have the advantage of being compulsory. Every fifteen-year old person should, and must, have an adequate knowledge in this matter. This is not to say that it is reasonable to attempt to demonstrate Bayes's theorem or the law of large numbers to an eleven-year old student, however, many activities, including entertaining ones, could be profitably carried out in such classrooms.

3 A Few Examples

It is customary, in primary schools, to ask pupils to complete the *house of 10*. There are several versions, and the one that could be called 'basic version' consists of the distribution of a printout in which, schematically, a small house is drawn, the 'roof' marked with 10, the 'side' divided in 9 pairs of squares – sometimes 10, though more rarely. The pupil's task is to fill the pairs with numbers whose sum is 10. Nothing wrong with the exercise, were it not for the fact that, once filled the pairs, the pupil thinks he has completed the work – since it not easy to avoid, at 7, to write the same pairs over and over again, while it is easy to forget some others.

Hence, the problem is that the pupil has been deprived of the possibility of discovering what he will later learn is called the *sample space*. Thus, it would be much better to let the pupils discover as many pairs as possible. In addition, once the search is over, they should be shown how to demonstrate, or at least to explain, within their intellectual capabilities, whether they have found all possible pairs, which leads to *generating criteria*. Such activities are certainly within the intellectual capabilities of such pupils, and contribute significantly to *combinatorial reasoning*.

To remain at the level of primary school, a pupil can be encouraged to develop the first mental images of *mathematical probability*. For instance, by starting to distinguish among events that are *impossible, probable* and *certain*. Among events that are *more probable* than others, to eventually reach the ability of finding the probability of *elementary events*.

At the level of junior high, it is possible to propose probabilistic situations of various kinds. In several games about throwing dies, double 6 is particularly desired, and the feeling is that this is a rare event: Is that true? Or isn't the outcome of the ordered pair (1;5) more or equally rare? Or of any other ordered pair?

On each student desk, a small opaque box can be placed that contains 2 red marbles and 2 blue ones. Before end of classes, one student gets hold of one marble, and the other student of one marble among the 3 remaining. Before leaving the classroom, the desk has to ordered according to the following procedure: If both marbles have the same color, student A has to do it, otherwise B does it. Who will, more often, be the first to get out?

Now, for both examples, how is it possible to compute event probabilities? Such examples, like all examples of this writing, are meant as illustration and can equally well substituted by others.

4 Further Statistics and Examples

Similar considerations can be made for *statistics*. Again, it would make little sense to talk about *variance* or of the *chi-square test,* but again many activities could be introduced, as the following examples illustrate.

According to some research carried out in a given region, children with large feet have better language skills than those with small feet. Does it follow, then, that children should be subject to physical exercises to enlarge their feet? Or, is it not more reasonable to conclude that feet grow together with the child, and that therefore a child with large feet is, almost certainly, older than one with small feet? Such a game could already be played at the kindergarten level.

In primary schools, it is possible to solve a game called *ciphered anecdote.* It can be solved in several ways, but knowledge of the frequency of alphabet letters in a given language helps considerably. After finding a solution in some way, it is possible to proceed with the method described by William Legrand in *The Gold-bug* by Edgar Allan Poe, which leads to a first, though rudimentary, statistical analysis of the language used.

Again for high school, the next is an example that, in the experience of the author, strikes people's fantasy. Consider the two diagrams

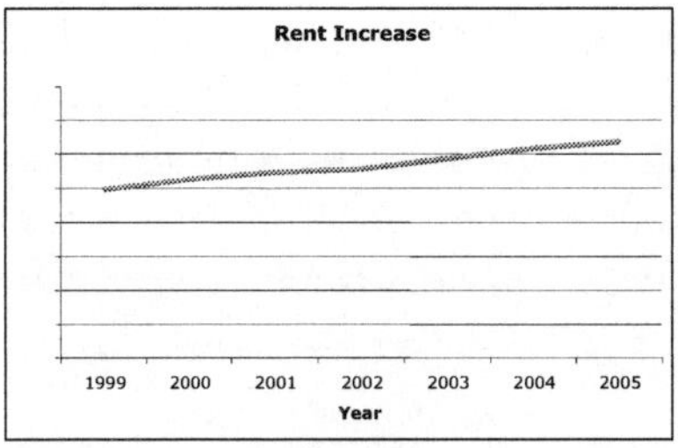

Diagram 1

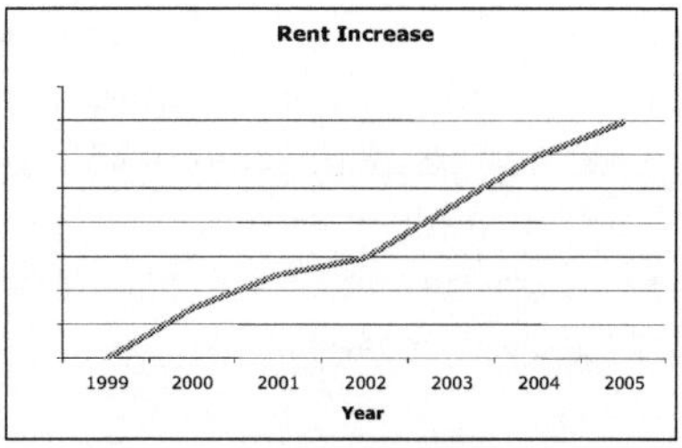

Diagram 2

and ask the question, which diagram shows the highest increase? Obviously the correct answer is that this can not be determined, since no values are given in the axis of the ordinates, however this is not the most frequent answer. Perhaps for this reason, most diagrams in newspapers do not have any values on the y-axis [...]. And the list could continue for as long as desired.

5 Conclusions: continuing teacher education

It is important that the idea of getting probability nearer to the students becomes the praxis of teachers. To need for notions of statistics is especially acute today, given the many abuses the public is subject to, and it is the task of the universities, the schools who educate such teachers, to spread such ideas.

But here lies the main difficulty. Not so much in the definition of curriculum of study in probability calculus – which are well established – but more so in the *education of teachers*, both fundamental and continuing, of a disciplinary character which is both historical and epistemological. To convince oneself that the problem lies not so much with students but with teachers, it is enough to enter a class and propose a stimulating probabilistic situation. It will soon be noticed that the students are capable of tackling such problems better than those of arithmetic and geometry – and this in front of surprised, sometimes even embarrassed, teachers.

However, teachers should not be penalized, because, in most cases, they have been abandoned. So-called updates, and continuing education, have been left to the good will of individuals, and this method can not guarantee an adequate preparation. It follows that there should not be any delusion about solving these problems with more or less excellent study curricula and splendid manuals. It is, instead, necessary to operate at the level of *teacher education*, *preparation* and *update of knowledge*.

Acknowledgement

The author wishes to thank colleagues and friends *Gianfranco Arrigo*, for the stimuli and the constructive criticism, and *Denis Baggi* for the translation in English of the original in Italian.

Giorgio Mainini, Teacher of Mathematics and Provost of Junior High School, Lugano

Sabrina Moretti

Chance and probability in bioinformatics

Abstract:

The discussion regarding the conceptions of chance and necessity has often been at the centre of attention for scientists. The greater emphasis ascribed to the role of chance in the life sciences, as opposed to physics, can be explained by the fact that the philosophical, ideological and, above all, religious implications of biology and biochemistry have always been very important.

Chance takes on a scientific value when it becomes possible to deal with it according to the laws of probability. But before using these laws, we need to understand the precise interpretation to be attributed to the concept of probability, in order that we can give it, at very least, an operative definition.

In the field of molecular biology, to take a particular example, the shift from the concept of chance *to the concept of* probability *has been realized by means of the definition of the informational metaphor. This has allowed us to perceive the living organism as being composed of a set of symbols, analysable through the use of frequentist statistics. In reality, this metaphor, although it has allowed us to obtain many important scientific results, does not allow us to solve all the problems presented. Specifically, it does not enable us to understand, in an exhaustive manner, the role of chance in scientific research; and some important questions remain unresolved, such as the relationship between microscopic and macroscopic levels, the attribution of meanings to genetic codes, and the control of complexity.*

Keywords:
probability, bioinformatics, genetic code, complexity, reductionism, symbolic code, information, frequentist statistics

1 From chance to probability

Reflection upon the concepts of chance and necessity has often been at the centre of attention among scientists. Naturally, biological sciences comprise many different fields, but when we speak of chance and necessity, in a biological field, we refer, above all, to the work of Jacques Monod [1970], who maintains that life is defined by these two factors:

> The final source of the project represented, pursued and realized by living beings reveals itself in this message [represented by a sequence of radical amino acids in a polypeptide filament], in this exact, faithful, but essentially indecipherable text constituted by its primary structure – indecipherable because, before expressing its physiologically necessary function, which it carries out in a spontaneous way, it reveals in its structure only the randomness of its origin [*id.*, p. 85].

Speaking about chance and necessity certainly had a very important cultural meaning when Monod published his book. Biologists have often discussed (and continue to discuss) the theme of 'chance', whereas other natural scientists deal with all non-deterministic phenomena by directly using the concept of probability. The greater emphasis ascribed to the role of chance in the life sciences – in comparison to physics, for example – can be explained by the fact that philosophical, ideological and, above all, religious implications of biology have always been very important. The importance of chance is directed towards defeating the last stronghold of vitalism, anti-Darwinism and biology's own ideological religious roots. For Monod, underlining the importance of chance in biology reflects an engagement which refers to the idea of a science that has a continuous need to free itself from various types of ideological conditioning.

However, as Boniolo [2004] points out, the concepts of chance and necessity become vague in an epistemological discussion. He maintains that chance and necessity do not exist, and that there are only random events and processes. An event is random if its achievement 'is bound to some causes that, for some reason, are undetermined, but which can be dealt with by using the laws of probability' [*id.*, p. 373]. The indeterminacy of causes may depend on various factors: it may be a problem of ignorance, when we believe that there are some causes but we do not have any instruments to discover them; or an ontological question; or perhaps an analytical inability due to the concomitance of several causes.

To deal scientifically with a random event therefore entails establishing an adequate interpretation of the concept of probability, and of the laws that we can apply to it. Currently, all theorized probabilistic models contemplate three possible interpretations: classic, frequentist and subjective. According to Boniolo, the most suitable interpretation for biology is the frequentist interpretation, even if it must be used with due caution, in order to avoid falling into the fallacy of undue generalization. Processes that occur within a cell are derived from complex interconnections among its components, and we can study them only in a probabilistic manner on the basis of empirical observations. The most suitable way to examine all intracellular processes is to use *a posteriori* probabilistic methods. To conclude, rather than speaking of chance and necessity, it would be more precise to refer to more or less probable events:

> It would be more correct to state that evolution takes place thanks to probabilistic mechanisms, which are quantifiable thanks to the theory of probability, having to do both with the appearance of new mutations and with genic rearrangements and chromosomal reabsorption. And these mechanisms create the matter from which the selection is then made. This selection is again a probabilistic mechanism whose higher or lower probability of incidence depends both on the type of selecting environment and on the particular correlation between this environment and the selected phenotype [*id.,* p. 42].

2 The informational metaphor

According to Boniolo, chance therefore assumes a scientific value when it becomes possible to deal with it by means of the laws of probability. Before using these laws, however, we need to understand which interpretation we should attribute to the concept of probability, in order to be able to provide it with an operative definition. The concept of probability does not make sense if it is not preceded by a vision of the world – that is to say, by an image (more or less precise) of the events and the objects whose nature we are exploring.

Biology arrives at an operative definition of probability when it accepts the reductionist model. Up until the latter half of the 19th century, biology was essentially descriptive, whereas in physics and chemistry the mechanist model was dominant, and the mathematical description was believed to be the bearer of absolute truth. Nature could be revealed by knowing all of its constituent parts, and all general mechanical rules governing it. The reductionist method is an essential element, and it is also mechanicism's greatest strength: the knowledge of any object can be reduced to a knowledge of its parts, and the whole is the sum of all of its components. Biology, by contrast, considered life to be qualitatively different from the rest of nature, and mechanicism and reductionism were therefore not suitable for studying it.

The mechanicist revolution in biology originated with a group of German physiologists, who claimed the universality of mechanicism and reductionism, maintaining that they could be applied outside the fields of physics and chemistry. This revolution subsequently spread to embryology, with the work of Roux and Mendel among others. However, the definitive turning point was in 1953, when Watson and Crick discovered the structure of DNA, following the intuition of Schrödinger, who had predicted the existence of a molecule in which were written all the elements characterizing human beings. The double helix model allowed scientists to explain the whole on the basis of a knowledge of its parts, exactly as reductionism requires. Mechanicist reductionism was to find its greatest expression in molecular biology, which achieved many successes during the latter half of 20th century.

Molecular biology also uses the 'informational' metaphor, according to which DNA is seen as a series of instructions – a genetic code. This metaphor developed in parallel with the computational sciences, and although biology and computer science are very different disciplines, they have often provided each other with sources of inspiration. In particular, the concepts of 'information' and 'program' have guided research in molecular biology. The genetic information is 'written' in DNA by means an 'alphabet' composed of four letters. It is first copied in RNA, with a very similar alphabet, and is then translated into the language, composed of twenty words, so to speak, of proteins, that are the essential elements for the functioning of the metabolism of living beings. The molecular structure of DNA and proteins can be represented by means of a codified message; the duplication of DNA and the synthesis of proteins can be considered as processes that transfer information, by means of special channels of communication, and as such, they can be dealt with by means of the same techniques developed in the theory of information. In brief, there is a sort of program written within the genetic code, which causes the functioning of all living organisms.

This metaphor has given rise to a new scientific discipline, bioinformatics, which acts as a bridge between biology and computer science. The concept of chance is interpreted in the light of Shannon's Theory of Information, which allows us to 'measure' the order of a sequence of symbols, by contrasting the concept of randomness with that of regularity. The techniques developed in the field of bioinformatics seek to find correlations and analogies between symbols of sequences, in order to measure the content and the entropy of information. Towards the end of the 1960s, the results of the Russian mathematician Andrei Nikolaevich Kolmogorov, well known for the notion of 'Kolmogorov complexity' in the field of probability theory, and Gregory Chaitin, the inventor of the algorithmic theory of information, contributed to the successful unravelling of the complex structures of sequences.

The fundamental objective of bioinformatics, apart from providing biologists with large amounts of experimental data, is to find patterns that emerge with a very low or very high frequency in DNA sequences. Biologists tend to attribute special significance to unusual patterns – that is, to patterns that have an anomalous frequency in comparison to others. This has motivated the development of statistical methods for the analysis of DNA sequences, with the aim of defining probabilistic criteria for discovering patterns with anomalous frequencies. It is of scientific interest, for example, to verify the presence of regularities in the four-letter symbolic structure of DNA. The presence of situations in which the distribution of these letters is not completely random can indicate a selective preference, derived from the adaptive capacity of a living organism.

Bioinformatics provides scientists with numerous statistico-computational techniques that may be applied to huge databases of 'sequenced' DNA, in order to search for statistical correlations among the letters to verify, for example, whether there is a causal relationship between the position of two letters in a sequence, or to calculate a degree of similarity between two or more sequences. Very similar sequences are much less informative than those with numerous points of divergence. This type of comparison allows the construction of phylogenetic trees that can help us to understand the processes of biological evolution.

3 Open problems

While the informational metaphor has allowed molecular biologists to obtain many useful results, it has also given rise to certain problems. The notion of a program has been transferred to biology only very roughly: it has been fruitful with regard to certain operative aspects, but has left many theoretical gaps. According to this metaphor, the DNA contains all the necessary instructions for creating the various components of a living organism, and these instructions can be broken up into numerous independent strings of information, the genes.

In recent years, in relation to the sequencing of the DNA of many living organisms, the deterministic genetic theory seems unable to explain how life works. A gene, for example, is a sequence of symbols that can be read in several different ways, and each type of reading leads to different meanings and thus to the synthesis of different proteins, each with a specific function. In addition to this ambiguity there are others, both during the transcription of the DNA 'message' and in the successive phases in which some proteins can configure themselves in various different ways in order to perform different functions:

> Here there arises the problem of understanding how variability, versatility and ambiguity are used, and therefore what causes the 'choice' of a cell to devote itself to this process or that one. First of all, we must understand the so-called problems of regulation with which a cell 'decides' to express a gene, and thus

> whether to produce a given protein or not; and we must also understand to what extent, and when and where (i.e., in which part of a multicellular organism), it should carry out this series of operations [Buiatti, 2003].

In short, we can say that each function of a living organism depends upon a combination of a great many events, involving DNA and proteins, and upon the correlations both within the DNA itself and between DNA and the proteins that form functional elements. Each function of the metabolism depends upon a series of steps that must be active 'in temporal harmony and with interdependent relative intensities' [Buiatti, 2000].

Buiatti maintains that the reductionist model is unable to provide explanations of the functioning of living organisms. The experimental data of recent years has, according to Buiatti, given us:

> a framework of living beings that differs greatly from that of the mechanical vision, even if it is based partly on data obtained during a period in which this vision was widely dominant in the scientific community. The novelty lies essentially in the discovery of the complexity and dynamics of living beings, and of the simultaneous presence of organization (order), on the one hand, and variability, versatility, ambiguity and redundancy (disorder), on the other; and all of these aspects are indispensable for maintaining order by means of change [Buiatti, 2003].

Life depends upon a complex system comprising proteins, enzymes, DNA and RNA, all of whose functions are correlated with one other in order to produce a highly sophisticated dynamic equilibrium. Each species has mutated over time by creating its own specific system of correlations. Living beings are provided with mechanisms that allow them to reach an equilibrium in response to internal or external perturbations, by readjusting their networks of correlations in such a way as to maintain the equilibrium of the system.

Living systems, therefore, should be examined in the light of the paradigm of complexity: the whole is not merely the sum of its parts, but something that emerges dynamically and in an unpredictable way from a network of correlations between its components, and from the relationship between order and disorder.

According to Sarkar [2005], the informational metaphor suggests an overly static image of the genome; it ignores the fact that the genome produces certain complex dynamics with regard to the conversion and transformation of DNA and proteins. In his opinion, such a metaphor should therefore be abandoned, and we should concentrate on the physical and chemical properties, not only of DNA but of the proteins synthesized within the cells. Sarkar maintains that in the future we shall need to pass from a genomic to a 'proteomic' viewpoint, reminiscent of the earliest spirit of molecular biology, when all types of molecule were the focus of interest, but proteins were viewed in a special way. Sarkar therefore holds that the reductionist model is valid, but that it needs to rid itself of the informational metaphor.

Atlan [1995] also asserts that the analogy between a genetic program and a computational program is a rather loose one. For one thing, a computational program, unlike a genetic program, is written by a programmer who has a certain purpose in mind. Further difficulties derive from the fact that even if DNA and proteins can be interpreted as messages carrying information, in reality we know only their symbolic code, and we ignore their programming language and the logical structure of the program: the syntax of the genetic code is unknown. It is therefore difficult to extend the notion of *genetic program* to the processes involving cells. According to Atlan, the absence of a programmer causes a theoretical difficulty that lies at the centre of our problems with the notion of a genetic program. This aspect has been ignored because biology was, until only a few years ago, based on a deterministic conception of the genetic code, and on the correspondence between genes and hereditary characteristics through the action of enzymes. Now, though, from observations on living organisms, scientists have ascertained that reality is much more complex.

Atlan maintains that we need to 'update' the analogy with a computational program by using more satisfactory models deriving from artificial intelligence and cybernetics, rather than from the older deterministic model. Moreover, these models arise with the aim of formalizing the concepts of complexity and self-organization that also interest biologists. The novelty is that they introduce the notion of a self-programming program. These instruments allow us to define self-organizing systems that, as a consequence of some aleatory perturbations and other randomizing factors, can redefine, in an autonomous manner, new properties capable of re-balancing a system. Self-organization implies a dynamic rearrangement of interconnected components, whereby the previous structure creates a situation conditioning this process, together with some random, unprogrammed perturbations that play a determining role. We find ourselves witnesses to a continual renewal whereby, when reacting to unpredictable and unprogrammed events, a system reorganizes itself instead of destroying itself.

Atlan adds, however, that this is not sufficient: it is necessary that for each new order a new significance be associated with the relationships that establish themselves among all the parts – that is, the new order needs to correspond to a new functional organization. Atlan writes:

> The most important aspect of the phenomena of self-organization is the self-creation of sense – that is, the creation of new meanings in the information transmitted from one part to another or from one level of organization to another level of organization. Without creation of new meanings, we would be dealing with recombinations that would not be able to give rise to new functions, new behaviours[…]. So that a disorganization is able to produce a reorganization, it is necessary to transform the meaning of the relationships among all the parts. This is why the issue of the creation of meanings is at the centre of all self-organization phenomena [*id.*, p. 167].

The idea, in essence, is to create new models that are able to modify themselves and to produce novelties that are unpredictable and surprising to an external observer.

Self-organizing models involve two factors: the role of chance in the evolution of the model, and the context in which the definition of meanings arises. The former produces unpredictable structures and behaviour, while the latter allows the attachment of meanings to these novelties, thus obtaining a new organization rather than simple chaos. 'In other words […] organization means allowing chaos to acquire a significance, inductively and in a certain context of observation' [*id.*, p. 168].

Therefore Atlan, like Buiatti, supposes the necessity of a theoretical turning-point that takes into account the complexity of this object, yet without abandoning the efficacy of the informational metaphor. Such a turning-point can be brought about by making use of new models derived from the most recent studies of artificial intelligence, the theory of automata and the theory of self-organization.

Barbieri [2003] maintains that an organic code is not just a metaphor, a linguistic construction useful in operative contexts, but a real entity. Codes generally establish a relationship between two symbolic systems. Genetic code sets up an association between codons (triplets of nucleotides) and amino acids. A 'true' code is characterized by two elements: it is arbitrary and it is defined by a 'codemaker'. Biologists consider organic codes simply as metaphors, because they think that they do not have these two features.

By using experimental data, Barbieri shows, first of all, the arbitrariness of genetic codes. Arbitrariness exists if there is no necessary connection between all symbols of code and what they represent. In the case of genetic code, we can demonstrate experimentally that nucleotides and amino acids are not linked by a chemical necessity. Each group composed of three nucleotides may be associated with any amino acid. 'Because it is not dictated by any chemical necessity, it can be realized only by way of an arbitrary choice, a real "natural agreement" ' [*id.*, p. 5].

A codemaker is the one who ascribes a significance to a code, by deciding how to establish a relationship between two symbolic systems. Our culture has accustomed us to thinking that only a mental or spiritual activity, related to human thought, can attribute this significance. According to Barbieri, meanings can also be established by natural, non-thinking entities; the important thing is that these relationships be created by a third entity, ontologically distinct from the two that determine the symbolic systems. We have always considered a cell as a dualistic system of genes and proteins, genotype and phenotype. According to Barbieri, we need to extend this conception to a triadic system, composed by genotype, phenotype and 'ribotype', where a ribotype is a system of ribonucleoproteins in a cell containing the molecules of the genetic code. The ribotype 'reads all instructions of the genotype and makes the proteins of the phenotype according to the rules of that code' [*id.*, p. 8]. The presence of ribonucleic protein systems suggests that genes and proteins are artefacts, artificial objects made by other natural entities. According to Barbieri, the subunits of molecules were originally ordered in a totally random way. He writes:

> At a certain point a new type of polymers appeared. Some molecules began to make some copies of some existing polymers[…]. Some other molecules began instead to make some polymers that were not direct copies but transformed copies – that is, codified versions of other polymers [*id.*, p. 9].

Thus 'copymakers' and 'codemakers' arise. Copymakers produce information, while codemakers produce 'meanings'. This hypothesis implies a new interpretation of natural evolution. Evolution does not derive solely from genetic mutations, but also from mutations of the genetic code – that is, of the instructions of the code – which occur in an independent manner. Barbieri concludes by saying that 'a convergent increase of complexity necessarily requires some memories and codes. In the organic world, in other words, there can be no epigenesis without organic memories and without organic codes [*id.*, p. 13].

To conclude, we can assert that the shift from the concept of *chance* to the concept of *probability*, in the field of molecular biology, has been realized through the definition of the informational metaphor. It has allowed biologists to perceive living organisms as being composed of a set of symbols that are analyzable by way of frequentist statistics. In reality, this metaphor, even though it has permitted the achievement of many extremely important scientific results, leaves certain problems unsolved. It is, for example, unable to explain the role of chance in a sufficiently exhaustive manner; and some important questions, such as the relationship between microscopic and macroscopic levels, the attribution of meanings to genetic codes, and the control of complexity, remain unresolved.

Bibliography

H. Atlan, 'Complessità, disordine e autocreazione del significato', in G. Bocchi & M. Cerreti (eds.), La *sfida della complessità*, Feltrinelli, Milan 1995.

M. Barbieri, 'I codici organici: metafore o realtà?', in *Sistema Naturae*, vol. 5, 2003, pp. 99-113.

G. Boniolo, 'Filosofia della biologia: che cos'è', in L. Floridi (ed.), *Linee di ricerca, SWIF* – Sito Web Italiano per la Filosofia (www.swif.it), 2004, <http://lgxserve.ciseca.uniba.it/lei/biblioteca/lr/public/boniolo-1.0.pdf>.

M. Buiatti, 'Prospettive darwiniane tra scienza e filosofia', paper presented at the International Summer School of Philosophy and History of Biology (Scuola Estiva Internazionale di Filosofia e Storia della Biologia), Fortezza Sangallo, Nettuno, Italy 2000.

M. Buiatti, 'Biologia moderna, biologia contemporanea, biotecnologie', in *Scienza esperienza*, 2003, <http://www.scienzaesperienza.it/news/new.php?id=0026>.

J. Monod, *Le hasard et la nécessité, Essai sur la philosophie naturelle de la biologie moderne*, Seuil, Paris 1970; It. transl., *Il caso e la necessità*, Arnoldo Mondatori, Milan 1970.

S. Sarkar, *Molecular Models of Life: Philosophical Papers on Molecular Biology*, The MIT Press, Cambridge MA 2005.

G. Valle, M. Helmer Citterich, M. Attimonelli, & G. Pesole, *Introduzione alla bioinformatica*, Zanichelli, Bologna 2003.

Sabrina Moretti, LCA – Laboratory for the Culture of the Artificial, Università di Urbino

Ephraim Nissan

Chance vs. causality, and a taxonomy of explanations

Abstract:

As Shafer has pointed out, chance is grounded in not knowing. Yet, it ought to be realized that abduction (i.e., inference to the best explanation) is based on realistic common sense only in a given kind of universe. The very fact that, e.g., mock-explanation is possible and is known to exist (e.g., in jokes, or in more complex literary works) shows that universes are conceivable, such that they obey laws different from the realistic common sense of the physical, social, and cultural world in which we function. Moreover, causality, and complementarily, chance, are differently ascribed, according to which account of the physical universe (or of the human world) we subscribe to. 'Realistic' explanation is better viewed as one kind, along with such explanations which are worldview-bound, or shaped by the poetic convention of a textual genre. It even happens that the distinction between bona fide and non bona fide explanation is irrelevant; we speak then of non-alethic (i.e., truth-independent) aetiologizing. One of the examples we develop, enables us to propose an original interpretation (itself a meta-explanation) of an aetiological tale from Pseudo Sirach.

Keywords:
Explanation (aetiology), explanatory purposes, causality and its ascription, chance, cultural relativity, narratives, art history, late-antique homiletics.

1 Accounts of the world

Ya'qūb ibn Iṣḥāq al-Kindī (يعقوب بن اسحاق الكندي) was a ninth-century philosopher, scientist, and physician (see Fig. 1).[1] Consider the physical conception of al-Kindi, or, more exactly, his conception in *On Rays* (as in fact he also held a different view on physics).

> The doctrine of rays of al-Kindi, the first Arabian philosopher, gives a generalised vision of a unique cosmic reality and a rationale for it. In that vision the four elements, bodies and planets make up a unified reality in which all parts seem to be made up by the very act of their constant relationship: every object projects its own nature outwards by means of rays which transmit that same nature to other objects. Thus, the relationship between the stars, the elements, and the bodies derive from and express the casual [*quod corrige:* causal] interactions which exist between these same levels of being, since only via this flux stars, elements and bodies can mutually connect. The cosmic unity is in constant flux due to the continuous transformation of everything, operated by the rays, and harmony is constantly attained through the recombining of forces coming from all parts of the cosmos.[2]

Fig. 1. A portrait of al-Kindi (not from his own lifetime)

Here human volition and agency come in:

> Thus 'magic' is practised on the basis of the interaction of all cosmic rays. This gives the 'magician', who aims at a specific purpose, the certainty of the effectiveness of his action, since 'magical power' acts within cosmic relationships, whose laws the 'magician' must understand in order to produce the desired effects. Indeed, the 'magician' 's intentionality itself acts as a ray.

Not only in the East, but in the West as well, some would accept al-Kindi's account of reality:[3]

> The medieval period, when the Kindian works were translated (the 12th century), lived a paradox of different and opposing conceptions of reality. Studying how and why some aspects of al-Kindi's thought were then appreciated or

> refuted, would possibly contribute to understanding how Kindian magic came over to become part of the 'dark side' of the Western tradition [...].

This is not to say that al-Kindi did not apply himself to conventional astronomy (and remarkably, he opposed astrology). In order to study the heavenly bodies, al-Kindi also concerned himself with, and wrote about, the armillary sphere, i.e., the spherical astrolabe, a model of the celestial sphere (with the Earth at the centre) invented by Eratosthenes of Cyrene in the third century BCE: see Celentano [1982]. See an armillary sphere in Fig. 2.

In Massimo Negrotti's naturoid-theoretic terminology,[4] an armillary sphere is a *concrete, analytical naturoid,* the observation level being planetary orbits, and the performance reproduced being positions in the respective orbits, relative to each other and to the centre. The model is planetary orbits, from nature. Moreover, because of the name of the device being derived from Latin *armilla,* i.e., 'bracelet', 'armlet', an armillary sphere is also a *concrete, aesthetic technoid,* the observation level being what is most conspicuous about the shape of the components of the device, and the performance reproduced being the circular shape of the rings. The model is a bracelet, which is a technological artefact.

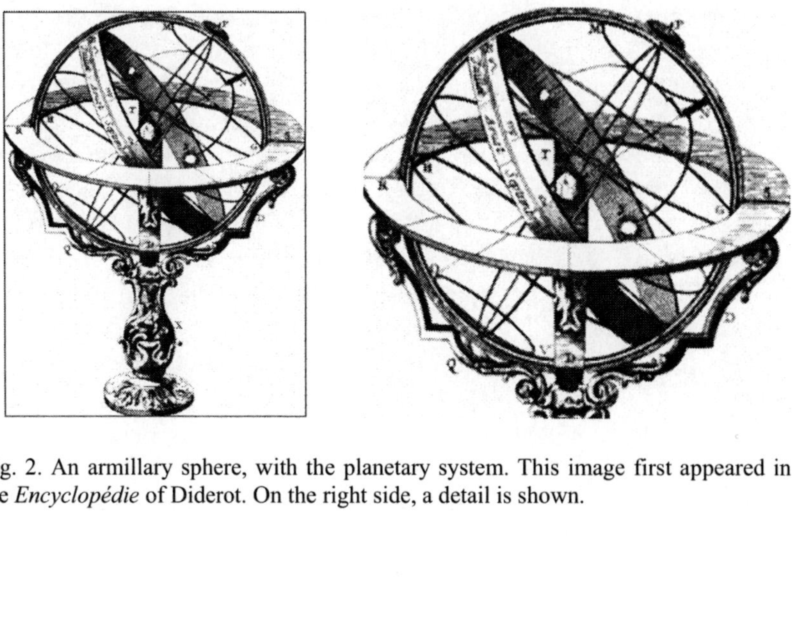

Fig. 2. An armillary sphere, with the planetary system. This image first appeared in the *Encyclopédie* of Diderot. On the right side, a detail is shown.

2 Culture-bound bodies of knowledge shaping explanations

If one does accept al-Kindi's conception concerning cosmic rays, then if, say, magician Tweety enacts a symbolic procedure to provoke the death of his foe, Sylvester, and Sylvester does die within a short time, then Tweety is liable for the death of Sylvester, even though Tweety did neither stab him, nor administered poison to him, nor falsely accused him, nor hired a hitman to do away with him.

Also, take the plague epidemic from the novel *I Promessi Sposi* by Alessandro Manzoni (1785-1873), set in the late 1620s and early 1630s. It was his only novel, yet it is the main novel in the Italian literary canon, and he is considered Italy's foremost novelist of the 19th century (the final version is from 1840). In Manzoni's novel, during the plague epidemic of 1629-1632,[5] the mob[6] blames innocents for propagating the contagion deliberately, whereas a minor character, the intellectual Don Ferrante, allegedly dies by blaming the planets (he believes in astrology). Both cognitive agents are keen on restricting the domain that a modern mentality would consider governed by chance.

Concerning mob's beliefs about the spreading of contagion in Manzoni's novel[7] (which is set in the early modern era), they found their most notable expression during the Black Death in the mid 14th century. That epidemic, the illness usually being identified as the bubonic plague – had reached western Europe in 1347, when twelve ships of Genoa reached the Sicilian port of Messina; they had arrived from the Black Sea, having left the Genoan-held town of Caffa in Crimea, which was besieged by the Mongols. The latter had catapulted, during the siege, the corpses of persons who had died of the plague inside the walls of the town. Historian Le Roy Ladurie [1973] memorably defined this the onset of 'the microbial unification of world'.

Significantly, a ritual murder tale was concocted, according to which, in Messina in 1347, Jews had crucified a Christian boy, a canard found, e.g., on pp. 469-470 in P. Samperi's *Iconologia* published in Messina in 1644, and still made much of (or rather re-launched) in a treatise published in Palermo in 1748, authored by Di Giovanni.[8] Cusumano [2004, p. 100], puts the date of Di Giovanni's treatise, 1748, in relation to the fact that in 1740 Charles III of Bourbon had decreed to readmit some Jews into Sicily (actually by 1748 such permission was already revoked, because of the protests the first decree had generated).[9] Even during the Black Death, accusing a minority of the contagion wasn't necessarily unchallenged: massacres of Jews took place, but in Strasbourg, where thousands were burnt alive, this didn't happen without the mayor proposing that an investigation should be carried out to find the actual causes of the epidemic; his motion wasn't carried.[10]

The tale of the Black Death in Sicily as related by Di Giovanni adds the element of blood, in a miracle allegedly accompanying the event, when as soon as the boy was thrown into a well, its water was turned into blood which then flooded the square. This putative concomitance of the epidemic with the fanciful murder story (which by the way, features prominently a well whose water was made undrinkable: a theme which had also been common in medieval Europe) shows a relation between accusations concerning the plague, accusations claiming ritual murder, as well as accusations of well-poisoning.[11]

In 1842, in his *Storia della colonna infame,*[12] Manzoni exposed the workings of the 17th-century myth of deliberate spread of the contagion.[13] Yet, even Manzoni himself could not bring himself to becoming associated with attempts to expose ritual murder libels,[14] possibly because official doxa were involved.[15] Within set parameters, such as a cardinal not discrediting a bull of the previous Pope,[16] one can see that attempts to contain and discredit the libel were made within the Curia.[17] Whether people would listen, is another matter. Sometimes when old doxa become unorthodox, some who considered themselves to be in line with orthodoxy would rather be unorthodox than change.

3 Causation, liability, and epistemology

It is also at the individual, autobiographical level, that sometimes a person would ascribe the cause of a negative situation to action by some perceived enemy, than to chance, or to his or her own errors. In his sympathetic portrayal of Mozart, Gutman [2001, p. 413] states that 'to his misfortune, he clung to his father's view of their disappointments not as functions of chance and blunder but, rather, of enemy conspiracy and perversity'. Such attitudes may be compounded by specific prejudice providing a ready-made explanatory pattern; Gutman [*ibid.,* p. 620], commenting on Mozart's early reaction to Da Ponte (a man far from spotless, yet his future librettist), remarks: 'Mozart had once again taken up the family practice of reading treachery in every Italian and conspiracy in every two'.[18] Significantly, Mozart's last decade was to see him succeed, and not unrelatedly, chart his course soundly and rationally, to a large extent emancipating himself from the behavioural and cognitive pitfalls of his early twenties. Explanatory fallacies affect the bulk of entire societies as well.

We have referred to the mob blaming supposed deliberate spreaders of the plague, and Don Ferrante blaming an astrological constellation, in Manzoni's novel. Consider, by contrast, the episode from Carlo Emilio Gadda's[19] famous novel (in an idiosyncratic Italian lexical pastiche), *Quer pasticciaccio brutto de Via Merulana* [*That Awful Mess on Via Merulana*] [Gadda 1957], which revolves around the investigation of a murder case in Rome. A chief inspector pays visit to a witch (and manager of a sweatshop) in the countryside, to find about a suspect. When he leaves, she places her hands on her belly, and moves her fingers while whispering. The policeman turns back, and orders her to stop it. Why? Is he insulted by her very intention? Or is he also superstitious, and is annoyed at the thought she may be successful at charming him? Readers with some knowledge of the early 20th-century society being portrayed[20] are left with little doubt that in all likelihood, the second explanation does apply indeed to this not unintelligent person of a rural background. Consider now the avowed skeptic or the closet Jacobin who may be watching that episode (something not contemplated by Gadda), and is a witness to an accident probably caused by the policeman being nervous, and in which the latter perishes (again, we, not Gadda, are making this up). Would that observer fail to find the coincidence striking? Would he or she blame the worthy lady for casting a spell, even though its only fairly ascribable effect was to make the policeman too nervous to drive? Is the witch liable, morally if not legally? If afterwards she is also run over by a car, would this be 'poetic justice'?

A different aspect of causality in respect of epistemology was analyzed by the philosopher Marshall Swain [1978].[21] Mr. Dunfor ingests a fatal dose of poison, then he is run over by a lorry and dies because of the latter event. This is complicated by Mr. Dunfor's also having ingested some food which along with the poison, resulted in a chemical reaction which neutralized the lethal effect of the poison, yet was also going to cause madness in Mr. Dunfor, such that he would have taken his own life, handn't it been the case that he was run over by a lorry. One's expectation he was going to die was based on his having ingested a fatal dose of poison, so it was unjustified.

We could turn, now, to the vast philosophical literature of epistemology, or, then, to computational logic-based theories of knowledge and action from artificial intelligence. Yet, let us probe the matter at a more basic and informal level.

4 Bona fide and non bona fide explanations other than realistic

The automated generation of explanations is an important topic within artificial intelligence. Inference engines for abductive reasoning have been developed. *Abduction* is defined as inference to the 'best' explanation. Refer to my review essay Nissan [2003]. One shortcoming of current research is the assumption (which for a vast category of applications is affordable) that abductive inference is to operate according to a set of commonsense rules, as though this was unique. In my ongoing GALLURA project, which has not entered as yet the computational design phase, I have been researching culture-bound devices of explanation, such as in aetiological (i.e., explanatory) tales from a given textual genre.[22]

It is necessary to distinguish between such explanations that are believed by the agent who provides them *(bona fide explanations)*, and such explanations that are not believed by the one who states them *(non bona fide explanations)*, either because this explanining agent intends to deceive the recipient of the explanation,[23] or because the explanation is humorous *(mock-aetiologizing)*, and the recipient can be expected to notice that this convention applies.

Both non bona fide and bona fide explanations may be based on either realistic commonsense (or current scientific views), or the rules of a universe which is not the universe of the late modern Western secularized worldview, as permeated by the current scientific vulgate. What one worldview would ascribe to chance, another worldview would ascribe to a causal link.

Moreover, also consider counterintuitive, yet realistic and bona fide explanations; consider the following example, from Venetian-ruled Crete.[24]

Fig. 3. A detail from a map from 1618 of Retimo in Crete. The fortress is bottom right.

① The town of Retimo[25] (Rethymni, Ρεθυμνη, in Greek)[26] used to have a fortress during the Renaissance (Fig. 3), but the city was poor and dilapidated.[27]

> Natale Dandolo capitano di Candia scrive nel 1580 che sarebbe stato meglio procedere alla distruzione della fortezza, possibile 'specchio' per i nemici, 'esca' così mal difesa per i Turchi [VENIVA 1996].
>
> Natale Dàndolo, the Captain of Candia, wrote in 1580 that it would have been better to demolish the fortress [of Retimo], as it was a possible attraction [literally: lark-mirror] for the enemy, and an ill-defended 'bite' that could bring in the Turks.

The proposition 'We ought to demolish that fortress of ours' is, on the face of it, incongruous, or at least counterintuitive, for the Venetian governor of Crete to put forth, yet the reason he gave was sound – even though it wasn't carried out, and that fortress was still standing in 1618, when Francesco Basilicata drew the map of the town and fortress ('CITTÀ ET FORTEZZA DI RETTIMO'),[28] an edited detail from which is shown in Fig. 3. The shape of the fortress from above is rather paw-like (perhaps like the footprint of a hedgehog).

② In Menicanti and Spiller's [1973] acclaimed dictionary (cum anecdotal repertoire) of the Milanese dialect, an entry renders Milanese *pianna* and *piannìn* respectively with Italian *pialla* ('[carpernter's] plane') and *piccola pialla* ('little plane'). Then an idiom is explained:

> *gh'è passaa dessôra San Giûsèpp cônt el piannìn,* l'è passato sopra San Giuseppe con la pialla. Si dice di una donna che non ha seno. Il detto era venuto di gran moda nel 1920 al primo apparire delle donne-crisi.
>
> St. Joseph planed her with his try plane. It is said of a woman who has no bosom. The idiom was quite popular in 1920, when the [fashion of the] crisis-women had just appeared.

This is a non-bona-fide, humourous aetiological tale. An anatomical feature is explained out as though it was the effect of finishing by means of an artisan's tool, as though the person was a piece of furniture. The implication is that the action was purposeful, and that this is how the end product was intended to look (which of course contrasts with the assumptions of both locutor and addressee). Moreover, the tale wouldn't do with just any carpenter; it had to be a very famous carpenter. This is an important difference with respect to realistic explanation, as in the latter perspective, pinpointing the identity of an unobserved agent is more costly that just ascribing the action to an unidentified member of an appropriate category of agents.

③ The explanation may concern the future, rather than reconstructing the past. In the following joke[29] in the form of a riddle, it is about mindreading (guessing other persons' thinking), only the reasoning ascribed is not usual common sense:

> Q. Betsie and Leo are marrying. Where should they go on honeymoon?
>
> A. To Madagascar. The Betsileo are the second largest population in that country.[30]

Examples of non bona fide explanation follow, based on an interpretation of an onomastic item.

④ Helmand is a river (600 miles long) of Afghanistan, flowing from hills below the Hindu Kush southwest to the Iranian frontier.[31] On 1 May 2006, *BBC Radio Four News* announced that over 3,000 British troops had been sent to the dangerous Helmand province in southern Afghanistan. Broadcasters kept uttering the toponym as though it was *Hell-manned* ['he:l̥mænd]. The context as conveyed was about a place replete with dangers, with Hell conditions, being manned. This may be a playful interpretation of the place-name, apt in the given context. You *don't* believe this is why the place got that name, yet you *do* believe in the aptness of the description.

⑤ In Geoffrey Treasure's *Who's Who in Early Hanoverian Britain* [Treasure 1992], the story is told, on pp. 231-232, of John Manners, Marquis of Granby (1721-1770): 'A lieutenant-general in Germany under Ferdinand, he had the mortification of sharing in the failure of the cavalry to complete the splendid victory of Minden. Sackville, commanding the cavalry, would not move; Granby, in command of the second line, was about to charge on his own initiative when Sackville rode up and ordered him not to. What the troops thought may be considered unprintable; what Ferdinand said was that, had Granby been in command, the French would have been annihilated. Sackville was dismissed and Granby succeeded him'. If more informal wording is allowed, you can replace *Sackville was dismissed* with *Sackville was sacked.* For this reason, Sackville turned out to be aptly named, by virtue of that particular episode.

⑥ In the same volume by Treasure, let us turn to passages dealing with another character from the same century [*ibid.,* p. 390]: 'SIR ELIJAH IMPEY (1732-1809) became Chief Justice of Bengal in 1774, earned the threat of impeachment from his association with Warren Hastings and the celebrated verdict of Macaulay: 'rich, quiet and infamous'.' In fact [*ibid.,* p. 391]: 'The alacrity with which Impey and his fellow judges were able to find Nand Kumar, the central figure in Francis's attack upon Hastings, guilty of forgery was also held against him. [...] So he came home in 1783 to find a menacing temper among the Whigs. In 1788 he was impeached and sat under a remarkable display of oratory by Elliot, who moved himself and others to tears with his elegiac upon 'the murdered rajah'. The impeachment was not pressed and he was acquitted honourably. However, he discovered that the popular view did not die so easily, when he stood for Parliament at Stafford in 1790. [...]'. 'The murdered rajah' was Maharajah Nandakumara, executed in 1775. These events are the topic of a book [Stephen 1885] by Sir James Fitzjames Stephen (1829-1894). Take the first and last words in the following sentence, from Treasure [*ibid.,* p. 400]: 'Impey, the Chief Justice who played an important part in Hastings's system, was dismissed and impeached'. Because of the similarity of the name *Impey* to the word *impeached,* Impey, too, appears to be quite aptly named, even though the impeachment against him was not pressed.

This kind of onomastic aptness is expressed, in traditional Jewish homiletics, by saying that the given character was so named 'because of his end', i.e., because of an event which eventually took place in that character's biography, well after the name was bestowed. Yet, epistemologically such a statement, if taken other than by way of literary convention of the given homiletical genre, is quite bold, and requires one to believe in forescience on the part of the name-giver. Such aetiological (i.e., explanatory) items blur the border between bona fide and non bona fide explanations; a cultural ideology which enabled much leeway in homiletics in order to expound a multifaceted sacred text for which truths may be concomitant,[32] would describe such explanations as bona fide. An example follows, from early rabbinic homiletics.[33]

⑦ A biblical event is reimagined according to how the name of a given character can be interpreted. Take King Nebuchadnezzar. The Book of Daniel relates that he threw into a furnace three of Daniel's companions who had refused to conform to idolatry, and that they were saved miraculously.[34] The moment when Nebuchadnezzar reacted angrily at their refusal, is reimagined in a homily by shaping the event both by reference to how his name, *Nebuchadnezzar,* could be segmented,[35] and arguably also by reference to another episode (when he lost his human reason and started to behave like an animal, only to be restored to reason and to his throne seven years afterwards: see Nissan [2007, fn. 7]).

In a passage in *Leviticus Rabbah* 33:6, the question is asked: why does the verse which relates the three's refusal, mention Nebuchadnezzar as 'King Nebuchadnezzar', whereas either 'the King' or 'Nebuchadnezzar' would have sufficed? The answer given is that the three companions (the ones who were to survive the fiery fire) had told him – so the homily maintains – that for matters like taxation, he is the King and they would comply with his order, but when it comes to matters of worship, he is Nebuchadnezzar, and when it comes to an order for them to become renegades to their faith, they would not comply any more than they would do, if faced with a dog ('you and a dog are equal'): he could as well 'bark *(neḇach)* like a dog, be blown up *(neṗach)* [distended] like a pitcher *(ḵad)*, and chirp *(neẓar)* like a cricket. He [Nebuchadnezzar] immediately barked like a dog, was blown up like a pitcher, and chirped like a cricket'. Which is how angry or almost apoplectic you could imagine a despotic king to become, on being challenged in the quite impertinent way described. (This way, the *omen* implied by the *nomen* became actualized once the person had been told how the name is to be interpreted.) Then, in the rest of the homily,[36] a verse from *Ecclesiastes* is cited in support of one having to obey a king, but not to such an extent that one would spite one's Creator. Arguably the reference to the King making sounds like an animal is an allusion to his period of madness: a story which is also impertinently embellished, in a different homily.

That this is not the most straightforward way to understand the primary narrative text (i.e., the narrative as told in the Bible) does not mean that the homiletic reinterpretation does not contribute any insight. Rather, this is an example of how human communication resorts to conventions other than realistic, possibly in order to convey some notion which may be relevant to real life (in this case, how far one can go in order to please political authorities, when the order is in contrast with the dictates of faith).

⑧ Let us consider an example that is non bona fide, more because of the character of the Hebrew-language source, Pseudo Sirach[37] (a popularistic, sometimes vulgar, text from outside religious literature, even though biblical themes occur there abundantly).[38] The frame story is about a child prodigy, Ben Sira, who answers questions asked by a rather domesticated King Nebuchadnezzar.[39] The medieval tales of Ben Sira merely outwardly adopt some strategies also found in the homiletical literature. Moreover, the given item rather resembles aetiological mythical tales as known from international folklore.

> And he [Nebuchadnezzar] also asked him [Ben Sira]: The bull, why does it have no hair under its snout?[40] He [Ben Sira] told him [the King]: When [the Children of] Israel and Joshua circumambulated Jericho, they would bring a donkey to Joshua, but it would collapse under him. Likewise, a horse, a mule, or any other beast would collapse, so heavy he was. Until they brought to him a bull. He mounted, and it was able to carry him. When Joshua saw this, he wanted to kiss it [i.e., the bull], and the bull raised its head, so Joshua kissed it under its snout. This is why it has no hair where it was kissed. He [Nebuchadnezzar] told him [Ben Sira]: Well done![41]

Even though this is a literary (however popularistic) text, it participates in the nature of folktales,[42] at least in the way it concocts its humour. The explanation is non bona fide. It is a mock-explanation. Yet, one is left to wonder whether there are, in the Levant (perhaps in Iranic myth?) any tales about a hero (or even pagan god) who is carried by a bull. One should not miss another factor, either: Joshua famously was a member of the tribe of Ephraim, and the bull is the biblical symbol of Joseph, from whom the tribes of Ephraim and Manasseh were descended. Therefore, Joshua is riding the emblematic animal of his tribe. Moreover, Islamic Arab legend, too, has the story about Joshua riding a bull,[43] but an explanation different from Joshua's being quite overweight is offered: it was Joshua who refused to ride on a horse, because horses had collaborated with the Pharaonic army during the pursuit into the Sea (where that army was drowned). This kind of aetiological variation is quite frequent in international folklore; being bona fide is not relevant: this is non-alethic (truth-independent) creativity. This is pertinent for mythical traditions.

5 When aetiological tales arise from a mythical tradition

Actually, it is possible to trace precedents in the mythologies of the Aegean, the Near East, Iranic lands, India, and Central Asia for bull riding. What has in all likelihood happened in *Pseudo Sirach* (see ⑧ in the previous section) is that a bull-riding hero, here, Joshua, was a given for the author. The traditional item (which we no longer possess) featured Joshua in association with the bull (this being the animal symbolizing his tribe, because it appears in Moses' blessing for [the descendants of] Joseph in *Deuteronomy* 33:17), and this may have been part of some no longer available (oral or written) source, yet one that also contributed to the Arab legend (perhaps originally Levantine Christian? or originally Islamic and Arabian?) about Joshua. Faced with the challenge of explaining Joshua's bull-riding, the author found a humorous explanation of why on earth should Joshua ride a head of cattle, rather than any equid (horse, mule, or donkey). As we are going to see, Iranian mythology has a particular king riding a bull, and a prominent narrative detail in *Pseudo Sirach* appears to originate in an Iranian myth:[44] after all, its author also tells a story about the child Ben Sira's birth, which has parallels in the mythology of Zoroastrianism (and also in Buddhism).[45]

In some myths, bull-riding is an indicator of strength: just consider, by way of comparison, bull-riding by cowboys. Because of the other sense of 'bull' in the American slang, i.e., the exchange going strongly, on the Web you can find an image on offer (with the smallest version free of charge), of a besuited businessman riding a charging 'bull' that actually is a skeleton made out of the symbol for the dollar (seen on 1 February 2007). See Figs. 4 to 6.

Fig. 4. The silhouette of a bull-riding cowboy.

Fig. 5. A businessman riding a bull-shaped dollar sign (edited from the original).

Fig. 6. More royalty-free clip-art: even a cactus can ride a bull.

In petroglyphs (i.e., rock carvings) from Tamgaly in Kazakhstan, there is a Bronze Age rather crude depiction of a man riding a bull, and acting as the Sun (it appears on a stamp of Kazakhstan from 2003, shown in Fig. 7):[46] of the man's body, only the barest outline is shown (an outline we usually associate with the drawings of children), but his head is a circle surrounded by rays.

Fig. 7. The stamp from Kazakhstan.

Archaeological excavations in Soli in Cilicia (i.e., coastal southeastern Anatolia) yielded an Astarte (a Near Eastern female deity) riding a bull, as reported at the end of the 19th century [AmJArchaeol 1899]. The Thessalian goddess Pheraia was depicted carrying a torch and riding a bull.

In Iranian mythology, King Ferêdûn (Freydun)[47] rides a bull while hunting, as well as his being the first to ride a bull at a given festival. This item from Iranian mythology, especially because of reference to a king (rather than a deity), arguably was a relevant possible influence, likelier than the rest, on the author of *Pseudo Sirach*. Importantly, for our hypothesis that the Joshua narrative was 'contaminated', in Mesopotamia, with the bull-riding feature of King Ferêdûn, note that during the Islamic era, in Islamic texts, Ferêdûn narratives and some biblical narratives were combined [Tafazzoli 2007]:

> Like other ancient Iranian heroes, Ferêdûn was identified in Islamic sources with such biblical and koranic figures as Noah [Bîrûnî, *Â*, p. 46], Abraham [Maqdesî, *Bad*ʿ, III, pp. 142, 144], and Nimrod [Dînavarî, ed. Guirgass, p. 8]. The introduction of Semitic elements into the legend of Ferêdûn also belongs to the Islamic period; for example, the story that he was twins in his mother's womb and of his miraculous birth [Maqdesî, *Bad*ʿ III, p. 142] is an adaptation of the story of Jacob and Esau.

All the more so, identification with Joshua is apt, as the bull is the symbol of Joshua's tribe, and cattle also features in the totemic ancestry of King Ferêdûn, which in turn prompted his association with cattle.[48] I would like to say in addition that beyond the Iranic area, in India, mythology also features bull-riding. One of the basic iconic features of the Tamil deity Civan (from southern India) is his riding a bull.[49]

Some other times, mythology has a character who is carried by a bull, but this is an associated with that character's weakness. This is certainly the case of the story, from Greek mythology, of the maiden Europa being kidnapped by Zeus, who appeared in the form of a bull and carried her away.[50] Yet, it has been suggested that originally, the character of Europa was a Cretan goddess who had lunar attributes, and was associated with the bull, a lunar animal (as being her 'associated consort animal', in technical parlance). If so, the myth of her being kidnapped may have developed in Classical Greece (or at any rate post-Minoan, post-Mycenaean Greece) as an explanation of her bull-riding iconography, an explanation which stemmed from a mentality which by then was reluctant to ascribe to women the stamina to ride a bull and still be in command. Again in Greek mythology, according to Nonnus, the youngster Ampelos was killed riding a bull maddened by Ate's gadfly. Ampelos was granted a second life, in the form of a vine, from which Dionysus squeezed the first wine. This, too, is an example of riding a bull, yet not being in control.

Fig. 8. The bull-riding hero from the Capitoline Museums. The relief appears on an altar base dedicated to Iuppiter Sol Serapis by Scipio Orphitus (inv.MC 1958/S). By kind permission of the Capitoline Museums in Rome. The relief is in Rome, at the Musei Capitolini; the photograph is at the Archivio Fotografico dei Musei Capitolini. The photograph (whose original is in colour) is part of the VRoma Project's Image Archive.[51] By kind permission of the photographer through the VRoma Project.[52] The photograph was taken in 2005 by Ann R. Raia. An edited version is shown here.

Perhaps the image that is most evocative of Joshua riding a bull, with a city (in his case, Jericho) in the background, is one that is preserved at the Capitoline Museums, at the Palazzo Nuovo in Rome. There are three reliefs from an altar with dedications to Jupiter, the Sun, and Serapis. One of the scenes in those reliefs (Fig. 8) shows a hero riding a bull beside goddess Italia, with a city background. The hypothesis of looking at mythology was warranted, as such a procedure is commonplace in the study of folktales (even such that appear in literary sources). Searching the Web provided some confirmation, by yielding much relevant material, which is summarized above. This way, we have devised a *meta-explanation* for the explanation in *Pseudo Sirach* about the bull, and Joshua riding that animal.

By being able to reconstruct, up to a point, the factors which militated for Joshua being made to ride a bull, and thereafter for the author of *Pseudo Sirach* meeting the challenge of explaining why Joshua did so, we have exemplified a reduction of the role of chance, through an augmentation of knowledge states: see in Section 9.

When it comes to a discussion of the devising, within ancient cultures, of (relatively) rational explanations for mythical tales, it is appropriate to mention the Greek author Palaephatus. He tried to rationalise tales from Greek mythology.[53]

6 Is it mere coincidence, or has the moon anything to do with it?

What is the role of chance in the following? We have seen that riding a bull has been associated, for the Cretan antecedent of Europa, with lunar attributes of both the bull (as a kind) and of the individual rider. The bull has no solar or lunar associations in the Hebrew Bible or later Jewish traditions, yet it is tantalizing that in *Pseudo Sirach,* bull-riding is stated for quite a specific event: the taking of Jericho. The name of Jericho, *Yeriḥó,* is certainly and quite transparently related to the Hebrew (and more generally Northwest Semitic) name for the moon, *yaréaḥ*. Is this just a mere coincidence? It could be by mere chance, yet one ought to consider: (a) whether there are any other lunar associates of Joshua; (b) whether the Iranic bull-riding mythical king had any lunar associates.

Let us start by arguing why Question (b) is relevant: Yassif [1984] remarked that *Pseudo Sirach* was likelier to have been influenced, when relating lore about King Solomon and the Queen of Sheba, by Islamic lore about them which was current in its author's days and place, than by less accessible, sporadic, related Jewish lore that did not pass through an Islamic conduit. This is strongly suggestive that

Pseudo Sirach may have been dependent on Islamic lore which conflated King Ferêdûn with some biblical character. Whereas, as seen earlier, extant sources variously associate Ferêdûn with Noah (in Bîrûnî), Abraham (in Maqdesî), and Nimrod (in Dînavarî), such variety suggest that further biblical identifications may have been current, and that *Pseudo Sirach* depended on an otherwise unknown tradition in which the conflation was with Joshua. This by itself does not answer Question (b), whether Ferêdûn had anything to do with the moon.

As to Question (a), the data are inconclusive: Joshua himself is the one for whose sake both the sun and the moon stop, or appear to stop, until (in a different episode) a given battle is over (*Joshua* 10:12). It would be perhaps befitting for Joshua, to whom both the sun and the moon deferred, to use a solar animal against a Moon City, but was there any perception of the bull as being solar in the given culture? Or then, should the bull be instead (like in ancient Crete) a lunar animal, would this be a case of homeopathic therapy, as expressed by the adage *similia similibus curantur*? Or at any rate of *recirculatio.*[54]

Apart from his association with Jericho, Joshua is only specifically associated with the moon as opposed to the sun in a single homiletical passage, as far as I know. That occurrence occurs in the *Babylonian Talmud* (hence, like *Pseudo Sirach,* from Mesopotamia, but Yassif denies in general that *Pseudo Sirach* depends on the Talmud); in tractate *Bava Bathra,* 75a, it is stated: 'The face of Moses is like the face of the Sun; the face of Joshua is like the face of the Moon'. This clearly reflects the epigonic, yet prominent status of Joshua, Moses' successor.

Remarkably, both Joshua and Moses share, instead, both water and telluric associates, the telluric (earth-related) facets being of the seismic kind for both Moses and Joshua. Joshua's patronymic is *bin Nun,* 'the son of Nun', which is interesting, because *nun* is Aramaic for 'fish'.[55] An aetiology of this patronymic – as thought it meant 'the son of a fish',[56] after the Aramaic sense of the word *nun* – motivated a legend about Joshua's birth and childhood.[57] Moses' association with water[58] is because of his being saved from the water while a baby, and because of his parting the waters of the sea.[59] In the homiletical tradition, even more relations of Moses to water can be found, even a suggestion that in order to ascend through the air to Heaven while on Mt. Sinai, he swam like a fish.[60]

As to Moses' telluric association, Mt. Sinai was shaking (*Exodus* 19:18) and smoking [*ibid.* and at 20:15], when the Ten Commandments were being given. Joshua's telluric association is found when he is being buried (*Joshua* 24:30) in his estate, in Mt. Ephraim (a region) and specifically north of Mt. Gaash *(har gá'ash)*. Homiletics noticed that *Mt. Gaash* occurs nowhere else in Scripture, and also noticed the homonymy of *gá'ash* with the Hebrew for 'quacking' (and when it indicates an emotion or mood: 'flaring up'); this was aetiologized accordingly in the *Midrash Shemuel,* 23:7: 'So what is *Mt. Gaash?* As the Children of Israel were negligent concerning him [i.e., Joshua] and did not eulogize him befittingly, the Lord was about having the mountain quake [also erupt?] on them'. Prominent medieval exegetes referred to this interpretation.[61] In Modern Hebrew, *har gá'ash* means 'vulcano', and the related verb means 'to erupt'.

All of the above suggest that Joshua himself was not felt to have any special relationship to the moon or the sun (to both, or to either). The lunar association of Jericho's is striking, though.[62] Within the main culture which is relevant for understanding *Pseudo Sirach,* the bull is the emblem of Joseph, thus, of Joshua's Ephraimite tribal identity. This is fairly adequate for explaining which connection inherent to Joshua made a bull a fitting animal with which to associate him, which *Pseudo Sirach* did. The Central Asian (and Iranic) prominence of Sun worship and the role hoofed animals had there in sun symbology suggest that, if Ferêdûn is involved in Joshua's bull-riding, this is with solar connotations, but perhaps the author of *Pseudo Sirach* would not have known that.

So is it by *sheer chance* that Joshua rides a bull, of all places, outside the walls of Jericho, whose name could be translated as *Moon City?* There is *a small chance* that this is not by chance. Or should I rather say, more negatively: *little chance.* Note the nuance.[63]

7 What if? Fitting data to extant knowledge

Yassif's [1984] critical edition of *Pseudo Sirach,* which he prefers to call *the medieval tales of Ben Sira,* is synoptic, i.e., the texts of two manuscripts appear on the same page. In the version whose background is in Italy, the tale about Joshua riding the bull is very similar to the one I translated (see ⑧ in Section 4), only the wording is somewhat different.

Let us do an exercise, or should I rather say, a *Gedankenexperiment.* Let us simulate the aetiologizing efforts of a cognitive agent from the Middle Ages. Suppose a medieval Italian Jew, who knew the Ben Sira tales (many certainly did), had come across a relief with the scene shown in Fig. 8 – which is extremely unlikely, but let us proceed nevertheless, for the sake of the attempt of simulation of how he would try to make sense of it. How would such a person match the event depicted, against his (or her, but more typically, his) cultural repertoire of known narratives?[64]

The bull-riding man 'must' be Joshua (whom else would our cognitive agent have ever heard about, as riding a bull?).[65] The city walls in the backdrop provide confirmation: they 'must' be those of Jericho (the ones which, circumambulated, would tumble down). The gate 'must' be shut close, as the primary, biblical narrative (i.e., the text of *Joshua,* 6:1) states this with a superlative: the two juxtaposed participles *sogéret u-mesuggéret* 'closing (f.sing.) and closed (f.sing.)'.

But how to explain the lady reclining on the ground? *Cherchez la femme* in the Jericho narrative. Arguably, the reclining lady is Rachab,[66] whose dwelling was in the walls of the city, and was spared (nor did her dwelling crumble along with the walls: at any rate, the people crammed inside were unhurt). Just perhaps. The problem is that in the relief, the woman is *already* in front of the man identified as Joshua, even though the city walls are standing, and therefore Jericho has not been taken yet.

One explanation that our hypothetical medieval viewer may come up with, is that by convention, the visual representation is free to cram in[67] different moments from the series of events (see Fig. 9).[68]

Fig. 9. Noah with the crow and the dove, from the Holkham Bible Picture Book. Courtesy of the British Library. The same picture shows different moments in the Noah narrative.

Thus, a medieval viewer would know, one picture is not necessarily to be considered as confined to representing just what we could call one 'time-slice' of a given story.[69]

An explanation of Fig. 8 which could be alternative to, or concomitant with, the explanation proposed earlier, would build upon the tradition according to which Joshua married Rachab,[70] which was (even though this is not stated explicitly) disregarding her at best equivocal professional vocation while still a pagan in Jericho.

Reference is appropriate to either interpretation of the descriptor, *zoná,* of Rachab's profession while in Jericho (*Joshua,* 2:2). The most straightforward is according to the usual sense of the term, 'prostitute'.[71] Mandelkern's biblical concordances describe her as 'meretrix Hierichuntina'. If so, then the depiction of the woman in the relief as being unclad may refer to her very coloured past, or should we say, her lamentable professional past. Our hypothetical medieval cognitive agent is less likely to retain this interpretive hypothesis. The likeliest explanation would be that Rachab is being shown unclad, as being a reference to the fact that her disavowal of paganism and recognition of monotheism (*Joshua,* 2:11) was followed by a formal conversion – whose most salient moment for a woman is taking a bath – thus confirming that her subsequent marriage to Joshua was quite kosher.[72] (Joshua's presence in the same scene confirms the relevance of such an interpretation: not just for identifying the circumstances of the event from sacred history of the taking of Jericho, but also, it would appear, as being a reference to his private life, i.e., to his marriage, about which the Bible says nothing.)

In case our cognitive agent is unwilling to admit that different moments of the story appear together in the same picture, then Rachab's presence outside the still standing walls of Jericho could be explained out by developing an aetiological tale: Rachab had been promised that anybody found inside her dwelling would not be harmed, yet she was anxious (and a confirmatory parallel can be found in Lot being promised that the little town of Zoar would not be destroyed along with Sodom and Gomorrah, so he could take refuge there, yet he is so anxious that instead of stopping in Zoar, he flees as far as the mountains). Being anxious (she had explicitly avowed her anxiety, in the biblical text), Rachab would have climbed down from her famous window (the way the explorers had done), just before the walls crumbled. Or, then, she may even have jumped down, which would befit extreme anxiety;[73] this would also explain out why she is reclining on the ground, and why she is partly undressed. This would have been an effect of the impact on her draped toga. To keep a toga in place, for a lady to use a fibula is functionally requisite,[74] but on impact, if the lady jumped out of the window, the fibula would not hold. ('So Rachab was wearing a toga? why not a tunic?' 'Of course she was wearing a toga; can't you see for yourself the sheet of cloth in the relief?').

What we are figuring out has no relevance for the philology of *Pseudo Sirach:* had a reader from the 11th century or later come across the relief from Fig. 8 (or across green men from Mars, for that matter), that would not have affected the text (unless the reader was going to write a new text inspired by the former). Yet, it says something about processes of weaving aetiological tales. Like late antique homiletics, medieval explanatory tales relating to either relics from history, or sacred texts, in a sense reflect an explanatory *horror vacui.* Our *Gedankenexperiment* of figuring out how a medieval viewer (aware of *Pseudo Sirach*), if faced with the scene from the Roman relief (Fig. 8), could have made sense of it, proceeded on the assumption of a reluctance to leave things unexplained. The relief had an author, who put the visual elements in the relief by design, not by chance. The medieval mentality of the viewer would try to read what was on the mind of the maker of that relief. Likewise, that same mentality would refrain from ascription to chance of this-worldly events, be they natural, or of international relations, or occurrences from everyday life, rather considering it a duty to acknowledge providential design, intervening both as general Providence, and, in the lives of individuals, particular Providence.

The cognitive simulation by which we derived alternative, branching aetiological subnarratives out of Fig. 8 also comes close to something which we do know had happened (among Christians visiting Rome or residing there) in the formation of a tale associating Constantine with an owl, based on a misunderstood detail in a sculpture of Marcus Aurelius. The next section deals with this.

(a)

(b)

Fig. 10. The sculpture of Marcus Aurelius from Capitol Square in Rome.[75]

(a) The copy of the original statue currently on display in Rome.[76]

(b) The same sculpture, on the Italian yellow coin of 50 cents of euro.[77]

8 The owl, and the plume of the emperor's horse

A detail in a famous sculpture from imperial Rome (Fig. 10), during the Middle Ages was interpreted as though it was a bird – an owl – which does not correspond to the fairly obvious present-day understanding of what that detail was originally intended to represent. Its interpretation as an owl gave rise to a narrative in which an owl could 'sensibly' fit. Refer to the text in emphasis towards the end of the following quotation. The context of the latter is about the medieval textual genre of the *Mirabilia urbis Romae,*[78] which catered to Catholic pilgrims, yet apart from sacred themes, also concerned itself with lay aspects of whatever remained to see, in Rome, of marvels of the pagan past.

> The *Mirabilia,* then, are first and foremost guides for the pilgrim to the places of worship, yet they of course also mention the remarkable profane things, the *marvels* of the pagan Rome; such parts draw our attention even more than the rest, because of the peculiarity of the exposition. The *Mirabilia,* indeed, belong to the same time and environment of the *Gesta Romanorum,* and resolve antiquity into a fable, similarly to what is to be found in that book of histori-

> cal-moral fables. It is the history of the world in antiquity being told to the children of the Barbarians in a quite evidently fantastical manner, just as the Roman poet from the Augustan era is transformed into the bizarre Virgil the magician and thaumaturgist. This fantastic and fabulous aspect of Italy is remarkably reflected in a letter sent to Hildesheim in 1194 by Henry VI's chancellor, Konrad of Querfurt. That same spirit is also to be found in the visual arts, especially in a set of profane ivory boxes, originating from Byzantium, as well as from the Venetian Laguna and from Norman Sicily. The strange allegorical stories from the *Mirabilia* which concern two of the most famous sculptures from ancient Rome, namely, the *Marcus Aurelius,* and the *Dioscuri* of Montecavallo, clearly display the spirit of this bizarre literature. The first, popularly named *Caballus Constantini* [i.e., Constantine's Horse], is the subject of the prodigious legend of the 'great peasant' [«gran villano»] and of the deliverance he brought to Rome. Even the mere fact is significant, that he was renamed Constantine and traces of this can even be found in the portals, decorated with sculptures, of churches in southern France. The extent to which the Middle Ages were in the habit of conceiving shape with simple naïveté and childish imagination, even on such a very ancient soil, can be seen in a little, yet essential detail, confirmed by parallels from the visual arts of that same period: **the plume of the emperor's horse is interpreted as though it was an owl**, thus providing yet another fabulous element. The medieval imaginary draws entirely fabulous stories from real works of art, as shown quite effectively by De Rossi concerning the well-known legend of the 'Clemence of Trajan', based on a Roman triumphal relief; [...].[79]

The ambiguity of the plume, as a visual element to be matched against the conceptual repertoire of the viewers, caused straightforward perceptual intepretation not to occur, and set the goal of categorizing the object. Had the plume been more amenable to categorization as something different than an owl, or had some other (not inconspicuous) part of the sculpture been more ambiguous, then narrative explanation would have taken a different direction. Such factors can be represented in terms of randomness. Computational techniques for pattern recognition can teach something about that. In a sense, based on what we have seen thus far, here we have a procedural outline of how chance can be involved in contexts which motivate the generation of aetiological tales.[80]

9 Chance and knowledge

In *The Art of Causal Conjecture,* as important a book as it is difficult, Glenn Shafer [2006][81] 'present[ed] a new mathematical and philosophical foundation for probability and show[ed] how this foundation can help us understand causality' (p. xv). In his approach in that book, 'causes can be represented as steps in a probability tree' (p. xvii). It was Huygens who first introduced the probability tree, in 1676, to organize probability calculations. Shafer adopts this concepts, and develops probability theory as based on causality, and as opposed to the sample-space approach, which has been usual in probability and statistics. Shafer accepts the view of probability as subjective certainty, as put forth by Jacob Bernoulli (1654-1704) in Part IV of his posthumous *Ars Conjectandi* [1713]. To Shafer [p. 1], contingency and probability

> are not subjective in the narrowest modern sense of the word; they do not pertain merely to a person's state of mind or disposition. But they are subjective in the broad sense in which Bernoulli used the word. They depend on the vantage point from which a person sees the world. They describe the person's knowledge and ability to predict. When I say it is contingent whether a coin will fall heads, I am acknowledging that I cannot calculate how it will fall. When I give a probability for its falling heads, I am expressing precisely my limited ability to predict how it will fall.

In fact [pp. 1-2]:

> Probabilistic ideas derive from a story about the relation between an observer and the world. As features of the observer's knowledge of the world, the probabilities in the story say something about the observer as well as something about the world. But in order to do justice to the objective as well as the subjective aspects of probability, we must stand back and recognize the diversity of ways the story can be used. [...] In particular, it can be used to describe how the world works.
>
> When we explain the working of the world in terms of objective probabilities, we are placing in the role of observer in the probability story an imaginary observer – we may call her nature – whose knowledge and prescience surpasses that of any actual observer but is noneless limited. Nature falls far short of God's omniscience, for she represents a hypothesized limit on the predictive ability that can be achieved by actual observers, human or artificial. Like actual observers, she watches events unfold step by step. Sometimes she

> can predict what will happen next, but more often she can only give probabilities.

'Chapter 14 develops the theme that causal explanations are partial descriptions of nature's tree' [ibid., p. 28]. We cannot even start, here, to review Shafer's approach to causality, as put forth in *The Art of Causal Conjecture.* Let us rather think of reconstructing a story. Consider the animal in Fig. 11(a). Was this a giraffe which was made to resemble a zebra? Or was it a zebra image like in Fig. 11(b), which was then warped into Fig. 11(a)? Readers are likelier to plump for the second, and correct, explanation. This is because we expect each other to possess socio-cultural knowledge about the world, including the distinction between an object and its representation – in this case, an aesthetic artificial (which reproduces appearance), in Massimo Negrotti's naturoid-theoretical terminology[82]– and we also know about graphic files: warping Fig. 11(b) into Fig. 11(a) is much easier than giving a giraffe a zebra's hide and other features while keeping the elongated neck.

Fig. 11. Understanding distortion.

(a) A zebra image with an elongated neck.

(b) The original zebra image.

That much realization will serve us in a companion article, scheduled to appear in the next volume of the *Yearbook of the Artificial,* 'On chance, causality, agency, volition, and communication: The case of the subversive parrot of Sfakion'. In the present article, we placed *explanation* at the centre. In the companion article, the centrepiece is *communication.*

Bibliography

A. Aarne and S. Thompson, *The Types of the Folktale: A Classification and Bibliography,* by A. Aarne, translated and enlarged by S. Thompson (Folklore Fellows Communications, 74). Suomalainen Tiedeakatemia = Academia Scientiarum Fennica, Helsinki, 1928. 2nd revision: (Folklore Fellows Communications, v. 75, no. 184), 1961. Reprints: 1973, 1964, 1981. Another reprint: B. Franklin, New York, 1971. Aarne's German original was *Verzeichnis der Märchentypen.* The latest addition is Uther (2004).

A. Adamatzky, *Dynamics of Crowd-Minds: Patterns of Irrationality in Emotions, Beliefs and Actions.* (World Scientific Series on Nonlinear Science, Series A, Vol. 54.) World Scientific, Singapore, 2005.

ALS News, Laser Time Slicing Promises Ultrafast Time Resolution. *ALS News,* 156 (July 12, 2000; last updated on September 30, 2002). Report posted at <http://www.als.lbl.gov/als/science/sci_archive/femto2.html> at *The Advanced Light Source* website (accessed in March 2007).

AmJArchaeol, 'Archaeological discussions: Summaries of original articles chiefly in recent periodicals', in the *American Journal of Archaeology,* 3(4/5), 1899, pp. 595-655.

C. Aslanov, *Le Provençal des Juifs et l'hébreu en Provence: le dictionnaire* Šaršot hakesef *de Joseph Caspi* (Collection de la Revuew des études juives, 21). Peeters, Paris & Louvain, 2001. [Based on MS Hebr. 1244 from the Bibliothèque Nationale de France.]

G. N. Atiye, *Al-Kindi, the Philosopher of the Arabs.* (In English and Arabic.) Kitab Bhavan, New Delhi, 1994; Islamic Research Institute, Rawalpindi, Pakistan, 1966.

S. Attardo, 'Trends in European humor research: Towards a text model', in *HUMOR: International Journal of Humor Research,* 1(4), 1988, pp. 349-369.

S. Attardo, 'Humor theory beyond jokes: The treatment of humorous texts at large', in: J. Hulstijn and A. Nijholt (eds.), *Automatic Interpretation and Generation of Verbal Humor: Proceedings of the 12th Twente Workshop on Language Technology,* Twente, 1996. University of Twente, Enschede, The Netherlands, pp. 87-101.

S. Attardo, Review of: G. Ritchie, *The Linguistic Analysis of Jokes* (Routledge, London, 2004), in *Pragmatics & Cognition,* 14(3), 2006, pp. 585-589.

S. Attardo and V. Raskin, 'Script theory revis(it)ed: Joke similarity and joke representation model', in *HUMOR: International Journal of Humor Research,* 4(3/4), 1991, pp. 293-347.

A. Aubel and D. Thalmann, 'Realistic deformation of human body shapes', in *Proceedings of Computer Animation and Simulation 2000,* Interlaken, Switzerland, 2000, pp. 125-135. Posted at <http://vrlab.epfl.ch/Publications> of the Virtual Reality Lab at the Swiss Federal Institute of Technology in Lausanne.

S. P. Baldwin and W. M. Saltzman, 'Polymers for tissue engineering', in *Trends in Polymer Science,* 4(6), 1996, pp. 177-182.

E. Ball, D. W. Chadwick and A. Basden, 'The implementation of a system for evaluating trust in a PKI environment', in: O. Petrovic, M. Ksela, M. Fallebblock and C. Kittl (eds.), *Trust in the Network Economy.* Springer-Verlag, Berlin, 2003, pp. 263-279.

A. Basden, E. Ball and D. W. Chadwick, 'Knowledge issues raised in modelling trust in a public key infrastructure', in *Expert Systems,* 18(5), 2001, pp. 233-249.

J. Bernoulli, *Ars Conjectandi.* Thurnisiorum, Basilea, 1713.

G. Bienaimé, *Moïse et le don de l'eau dans la tradition juive ancienne: Targum et Midrash* (Analecta Biblica, 98). Biblical Institute Press, Rome, 1983. Based on a dissertation: Institut biblique pontifical de Rome, 1981.

G. Bos and C. Burnett, *Scientific Weather Forecasting in the Middle Ages. The Writings of al-Kindi: Studies, Editions, and Translations of the Arabic, Hebrew and Latin Texts* (in English, Arabic, Hebrew and Latin), in the Sir Henry Wellcome Asian series, Kegan Paul International, London, 2000.

Brepols, *Medieval and Renaissance Studies Newsletter,* 20, Brepols, Turnhout (Belgium), Sept. 2003.

K. Čapek, *R.U.R.: Rossum's Universal Robots* [a play]. Aventinum, Prague, 1921. English editions: transl. B. L. Appleby, Lancaster University, 1970; transl. P. Selver, Oxford University Press, Oxford, U.K., 1961, 1991.

V. Carofiglio, V., & de Rosis, F., 'Exploiting uncertainty and incomplete knowledge in deceptive argumentation', in *Computational Science, ICCS 2001* (Lecture Notes in Computer Science, 2073). Springer-Verlag, Berlin, 2001, pp. 1019-1028.

V. Carofiglio, F. de Rosis and R. Grassano, 'An interactive system for generating arguments in deceptive communication', in: F. Esposito (ed.), *Proceedings of AI*IA 2001: Advances in Artificial Intelligence* (Springer Lecture Notes in Artificial Intelligence, 2175). Springer-Verlag, Berlin, 2001, pp. 255-266.

R. Cassinis, L. M. Morelli and E. Nissan, 'Emulation of human feelings and behaviours in an animated artwork', in the *International Journal on Artificial Intelligence Tools,* 16(2), pp. 291-375, 2007 (85 pp.). Full-page table of contents of the article on p. 158.

C. Castelfranchi and I. Poggi, *Bugie, finzioni e sotterfugi.* Carocci, Florence, 1998.

C. Castelfranchi and Y. Tan, *Trust and Deception in Virtual Societies.* Kluwer, Dordrecht, The Netherlands, 2002.

L. de Cataldo Neuburger and G. Gulotta, *Trattato della menzogna e dell'inganno.* Giuffrè, Milan, 1996.

G. Celentano, *L'epistola di al-Kindi sulla sfera armillare.* (In Italian and Arabic.) Supplemento n. 33 agli *Annali,* 42(4). Istituto Orientale di Napoli, Naples, 1982.

D. W. Chadwick and A. Basden, 'Evaluating trust in a public key certification authority', in *Computers and Security,* 20(7), 2001, pp. 592-611.

D. W. Chadwick, A. Basden, J. Evans and A. J. Young, 'Intelligent computation of trust', short paper at *TNC'98,* 1998.

B. Chiesa, *Creazione e caduta dell'uomo nell'esegesi giudeo-araba medievale.* Paideia, Brescia, 1989.

E. Cosquin, 'Le lait de la mere et le coffre flotant', in *Revue des Questions Historiques,* 83, 1908, pp. 353-425.

C. Crittendon, *Unreality: The Metaphysics of Fictional Objects.* Cornell University Press, Ithaca, Ney York, 1991.

N. Cusumano, 'Ricerca sull'accusa di omicidio rituale nel Settecento', in *Mediterranea: Ricerche storiche,* 1(2), 2004, pp. 81-104. (Posted on the Web at the address: <www.ddev.it/public/md1_dir/r67.pdf>).

L. Dégh and A. Vázsonyi, 'The memorate and the proto-memorate', in *Journal of American Folklore,* 87, 1974, pp. 225-238.

A. Demirhan Erdemir, 'A perspective of an English translation of *Aqrabadhin* of Abu Yusuf Ya'qub ibn Ishaq al-Kindi and some results', in *Journal of the International Society for the History of Islamic Medicine (ISHIM)*, 1(1), 2002, pp. 16-20.

H. Dieckmann and J. Seznec, 'The horse of Marcus Aurelius: A controversy between Diderot and Falconet', in the *Journal of the Warburg and Courtauld Institutes,* London, 15(3/4), 1952, pp. 198-228.

M. Dor, *Ha-ḥay bi-ymei ha-Miqra ha-Mishnah ve-ha-Talmud* [The Fauna at the Times of the Bible, the Mishnah and the Talmud] Grafor-Daftal Books, Tel-Aviv, 1997.

M. G. Dyer, *In-Depth Understanding: A Computer Model of Integrated Processing of Narrative Comprehension.* The MIT Press, Cambridge, Mass., 1983.

L. Edmunds, *Oedipus: The Ancient Legend and Its Later Analogues.* Johns Hopkins University Press, Baltimore, 1985.

Y. Elstein, 'The Gregorius legend: Its Christian versions and its metamorphosis in the Hasidic tale', *Fabula,* 27, 1986, pp 195-215.

A. Esposito and D. Quaglioni, *Processi contro gli ebrei di Trento (1475-1478).* Vol. 1: *I processi del 1475.* (In Italian, with Latin documents.) CEDAM, Padova (for the Dipartimento di Scienze Giuridiche of the Università di Trento), 1990.

A. Fassò, 'La duplicità dell'indovinello veronese', in *Rivista Italiana di Dialettologia,* 17 (1993), pp. 25-53.

P. Fehl, 'The Placement of the Equestrian Statue of Marcus Aurelius in the Middle Ages', in the *Journal of the Warburg and Courtauld Institutes,* London, 37, 1974, pp. 362-367.

A. FitzLyon, *The Libertine Librettist: A Biography of Mozart's Librettist Lorenzo da Ponte.* John Calder, London, and Abelard Schumann, New York, 1955. Republished as *Lorenzo da Ponte: A Biography of Mozart's Librettist.* John Calder, London, and Riverrun Press, New York, 1982.

G. Flügel, *Al-Kindî genannt der Philosoph der Araber: Ein Vorbild seiner Zeit und seines Volkes,* in the series Abhandlungen für die Kunde des Morgenlandes, 1(2). Brockhaus, Leipzig, 1857.

G. Fritz, Review of Verhagen (2005), in *Pragmatics & Cognition,* 14(3), 2006, pp. 589-597.

C. E. Gadda, *Quer pasticciaccio brutto de via Merulana,*[83] Garzanti, Milan, 1957. English translations: *That Awful Mess on Via Merulana*, translated by W. Weaver. Braziller, New York, 1966. Secker & Warburg, London, 1965.[84] Published again: *That Awful Mess on Via Merulana* <http://www.arts.ed.ac.uk-/italian/gadda/Pages/resources/babelgadda/babeng/calvinointromess.html> (with an introduction by I. Calvino), translated by W. Weaver. New York: Braziller, New York, 1985, and Quartet Encounters, London, 1984. A partial translation: *A New Annotated Translation of C. E. Gadda's 'Quer pasticciaccio brutto de via Merulana',* translated by R. de Lucca. In *Forum Italicum* 34(1), 2000, pp. 270-286.[85]

I. Garro, 'Al-Kindi and Mathematical Logic', in *International Logic Review,* 17/18, 1978, pp. 145-149.

I. Garro, 'The Paradox of the Infinite by al-Kindi', in the *Journal for the History of Arabic Science,* 10(1/2), 1994, pp. 111-118, 143.

V. Giura, 'Gli Ebrei e la ripresa economica del Regno di Napoli 1740-1747', in: V. Giura (ed.), *Storie di minoranze: Ebrei, Greci, Albanesi, nel Regno di Napoli.* Edizioni Scientifiche Italiane, Naples, 1984.

F. Grasso, 'Would I lie to you? Fairness and deception in rhetorical dialogues', in: R. Falcone and L. Korba (eds.), *Working Notes of the AAMAS 2002 Workshop on 'Deception, Fraud and Trust in Agent Societies',* Bologna, Italy, 15 July 2002 (2002a). (The article is posted on the Web, and can be downloaded from <http://www.csc.liv.ac.uk/~floriana/pub.html>).

F. Grasso, 'Towards computational rhetoric', in *Informal Logic Journal,* 22(3), 2002b, pp. 225-259.

R. R. Guerrero, *La recepción árabe del 'De anima' de Aristoteles: Al-Kindi y Al-Farabi,* in the series Colección Pensamiento islámico, 1. Consejo Superior de Investigaciones Científicas, Madrid, 1992.

M. Gutierrez, D. Thalmann, F. Vexo, L. Moccozet, N. Magnenat-Thalmann, M. Mortara and M. Spagnuolo, 'An ontology of virtual humans: Incorporating semantics into human shapes', in *Proceedings of the Workshop towards Semantic Virtual Environments (SVE05),* Villars, Switzerland, March 2005, pp. 57-67. Posted at <http://vrlab.epfl.ch/Publications> of the Virtual Reality Lab at the Swiss Federal Institute of Technology in Lausanne.

R. W. Gutman, *Mozart: A Cultural Biography.* Pimlico (Random House), London, 2001. Previously: Harcourt Brace & Co., New York, 1999; Secker & Warburg, London, 2000.

P. Hardie, Review of: J. J. O'Hara, *True Names: Vergil and the Alexandrian Tradition of Etymological Wordplay* (University of Michigan Press, Ann Arbor, 1996). In *International Journal of the Classical Tradition,* 6(2), 1999, pp. 284-286.

P. J. Hayes, 'Naive Physics I: Ontology for liquids'. Ch. 3 in: J. R. Hobbs and R. C. Moore (eds.), *Formal Theories of the Commonsense World.* Ablex, Norwood, New Jersey, 1985.

A. M. Heindorff, 'Petroglyphs within the Archaeological Landscape of Tamgaly (2004), Kazakhstan', posted on the Web at the following address: <http://worldheritage.heindorffhus.dk/frame-KazakhstanTamgaly.htm> (it had been last revised on 21 July 2006, when accessed on 1 February 2007).

B. Heller, 'Yusha bin Nun', *The Encyclopaedia of Islam,* 1st edition, vol. 4, 1934, 1177.

B. Heller, 'Yūnus b. Mattai', in English, in: H. A. R. Gibb and J. H. Kramers (eds.), *Shorter Encyclopaedia of Islam,* Brill, Leiden, 1953, repr. 1974, pp. 645-646. Originally in *The Encyclopaedia of Islam,* 1st edn.

B. F. Herring, *Joseph ben Kaspi's* Gevia' kesef: *A Study in Medieval Jewish Philosophic Bible Commentary.* New York: Ktav Pub. House, 1982. [Includes Caspi's Hebrew text and an English trans.; based on a PhD dissertation, Yeshiva University, New York, 1976.]

S. Hodges, *Lorenzo da Ponte: The Life and Times of Mozart's Librettist.* Granada, London, 1985.

A. Holden, *The Man Who Wrote Mozart: The Extraordinary Life of Lorenzo Da Ponte.* Weidenfeld & Nicolson, London, 2006; Phoenix (Orion Books), London, 2007.

L. Honko, 'Memorates and the study of folk beliefs', in *Journal of the Folklore Institute,* 1, 1964, pp. 5-19.

S. Horovitz and I. A. Rabin, *Mechilta d'Rabbi Ismael* (in Hebrew). Bamberger & Vahrman, Jerusalem, 5720 = 1960.

R. P.-C. Hsia, *Trent 1475: Stories of a Ritual Murder Trial.* Yale University Press, New Haven, Connecticut (with Yeshiva University Library, New York), 1992; new edition: 1996.

I. Infantino, R. Cipolla and A. Chella, 'Reconstruction of architectural scenes from uncalibrated photos and maps', in *Proceedings of IAPR Workshop on Machine Vision Applications,* Tokyo, Japan, November 2000.

H. Jason, *Folktales of the Jews of Iraq: Tale-Types and Genres.* In English and Hebrew. Babylonian Jewry Heritage Center, Or Yehuda, Israel, 1988.

J. Jolivet, *L'intellect selon Kindi.* (Publication de la Foundation De Goeje, 22.) Brill, Leiden, 1971. (It includes Arabic text and a French translation of al-Kindi's *Risālat al-'aql*).

J. Jolivet and R. Rashed, 'al-Kindī, Abū Yūsuf Ya'ḳūb b. Isḥāḳ', in English, in *The Encyclopaedia of Islam, New Edition,* vol. 5, Brill, Leiden, 1986, pp. 122-123.

J. Kadane and D. Schum, *A Probabilistic Analysis of the Sacco and Vanzetti Evidence.* Wiley, New York & Chichester, England, 1996.

M. Kadushin, *A Conceptual Approach to the Mekilta.* New York: Jewish Theological Seminary of America, New York, 1969 [1981]. (Includes vol. 1 of Lauterbach's edition and translation.)

O. Kahl, *Sābūr ibn Sahl: Dispensatorium parvum (al-Aqrābādhīn al-ṣaghīr).* Brill, Leiden, 1994.

R. Kushelevsky, *Moses and the Angel of Death* (Studies on Themes and Motifs in Literature, 4.) Peter Lang, New York and Bern, 1995.

I. La Lumia, 'Gli ebrei siciliani', in his *Studi di storia siciliana.* Virzi, Palermo, 1883. Published again (introd. F. Giunta): Edizioni della Regione Siciliana, Palermo, 1969, Vol. 2.

P. V. Lamarque and S. H. Olsen, *Truth, Fiction, and Literature: A Philosophical Perspective.* Clarendon Press, Oxford, 1994.

A. Lanapoppi, *Realtà e leggenda nella vita del librettista di Mozart.* Il Mulino, Bologna, 1992.

J. Z. Lauterbach, *Mekilta de-Rabbi Ishmael: A critical edition on the basis of manuscripts and early editions with an English translation, introduction and notes* (3 vols.; The Schiff Library of Jewish Classics). Jewish Publication Society of America, Philadelphia, 1933-1935 (subsequently, a facsimile reprint).

E. Le Roy Ladurie, 'Un concept: l'unification microbienne du monde (XIVe-XVIIe siècles)', in *Revue Suisse d'Histoire,* 23, 1973, pp. 627-696.

M. Levey (ed., trans.), *The Medical Formulary of Aqrābādhīn of al-Kindī,* facsimile text, with English translation and notes. University of Wisconsin Press, Madison, 1966.

B. Lincoln, *Priests, Warriors and Cattle.* California University Press, Berkeley, 1981.

D. C. Lindberg, *Theories of Vision from al-Kindi to Kepler,* in the series The Chicago History of Science and Medicine, University of Chicago Press, Chicago, 1976, paperback edition.

S. Mandelkern, *Veteris Testamenti Concordantiae Hebraicae atque Chaldaicae,* 1896 (with definitions in Latin, and annotations in Hebrew). Edition edited by F. Margolin and M. Goshen-Gottstein, Schocken, Jerusalem, 1977.

A. Marchesan, *Della vita e delle opere di Lorenzo Da Ponte.* Turazza, Treviso, 1900.

S. Matton (introd.), *La Magie arabe traditionnelle,* in the series Bibliotheca hermetica, Retz, Paris, 1976.

S. Menicanti and A. Spiller, *Vocabolario del milanese d'oggi: repertorio storico, etimologico e aneddotico.* Rizzoli, Milano, 1973.

B. Mesch, *Studies in Joseph Ibn Caspi: Fourteenth-Century Philosopher and Exegete* (Études sur le judaisme médiéval, 8), Brill, Leiden, 1975. Based on a thesis, Brandeis University, Waltham, Massachusetts.

M. Moreshet, *A Lexicon of the New Verbs in Tannaitic Hebrew* (in Hebrew). Bar-Ilan University Press, Ramat-Gan, Israel, 1980.

M. Negrotti, 'Towards a general theory of the artificial', in: M. Negrotti (ed.), *The Culture of the Artificial,* special issue, in *AI & Society,* 14(3/4), 2000, pp. 268-299.

M. Negrotti, *Naturoids: On the Nature of the Artificial.* World Scientific, Singapore, 2002.

J. Neusner, *Mekhilta According to Rabbi Ishmael: An Introduction to Judaism's First Scriptural Encyclopaedia.* (Brown Judaic Studies, 152). Scholars Press, Atlanta, Georgia, 1988a.

J. Neusner, *Mekhilta According to Rabbi Ishmael: An Analytical Translation.* (2 vols.; Brown Judaic studies, 148 & 154). Scholars Press, Atlanta, Georgia, 1988b

J. Neusner, *The Canonical History of Ideas. The Place of the So-Called Tannaitic Midrashim: Mekhilta Attributed to R. Ishmael, Sifra, Sifre to Numbers, and Sifre to Deuteronomy.* (South Florida Studies in the History of Judaism, 4). Scholars Press, Atlanta, Georgia, 1990.

F. M. Nichols (trans., ed.), *The Marvels of Rome, or, A Picture of the Golden City: An English Version of the Medieval Guide Book.* Ellis and Elvey, London, 1889.

E. Nissan, 'Culture-bound technological solutions: An artificial-theoretic insight', in: M. Negrotti (ed.), *The Culture of the Artificial,* special issue, in *AI & Society,* 14(3/4), 2000, pp. 411-439.

E. Nissan, 'Facets of abductive reasoning'. [Review essay on: L. Magnani, *Abduction, Reason, and Science: Processes of Discovery and Explanation* (Kluwer / Plenum, New York, 2001); J. R. Josephson and S. G. Josephson (eds.), *Abductive Inference: Computation, Philosophy, Technology* (Cambridge University Press, Cambridge, England, 1994); H. Bunt and W. Black (eds.), *Abduction, Belief and Context in Dialogue: Studies in Computational Pragmatics* (Benjamins, Amsterdam, 2000).] In *Cybernetics and Systems,* 34(4/5), 2003, pp. 381-399.

E. Nissan, 'A gleaning of concepts from the natural sciences held by the Jewish sages of late antiquity: From zoology, to optical instrumentation (viewing tubes)', in press in the proceedings of the *XXXV Incontro di Studiosi dell'Antichità Cristiana,* held on 4-6 May 2006 in Rome at the Augustinianum: Istituto Patristico, on the theme *La cultura scientifico-naturalistica nei Padri della Chiesa (I-V sec.).* Pontificia Universitas Lateranensis, Rome.

J. J. O'Connor and E. F. Robertson, 'Abu Yusuf Yaqub ibn Ishaq al-Sabbah Al-Kindi' (in English), School of Mathematics and Statistics, University of St. Andrews, Scotland, 1999. Posted on the Web at the address: <http://www-history.mcs.st-andrews.ac.uk/Biographies/Al-Kindi.html> (accessed in Jan. 2001).

D. Park, *The Fire Within the Eye: A Historical Essay on the Nature and Meaning of Light.* Princeton University Press, Princeton, New Jersey, 1997.

R. Pettazzoni, 'The pagan origins of the three-headed representation of the Christian Trinity', in the *Journal of the Warburg and Courtauld Institutes,* London, 9, 1946, pp. 135-151.

D. E. B. Pollard, 'Logic of fiction' in: P. V. Lamarque and R. E. Asher (eds.), *Concise Encyclopedia of Philosophy of Language.* Pergamon, Oxford, 1997, pp. 264-265.

G. Priest, *In Contradiction: A Study of the Transconsistent.* Martinus Nijhoff, Dordrecht, The Netherlands, 1987.

G. Priest, 'Paraconsistent Logic', in *Handbook of Philosophical Logic,* 2nd edition, Vol. 6, ed. by D. Gabbay and F. Guenthner. Kluwer, Dordrecht, 2002, pp. 287-393.

G. Priest and K. Tanaka, 'Paraconsistent logic', in the *Stanford Encyclopedia of Philosophy* (online on the Web), 1996, 2000, revised 2007 (accessed in 2004 and 2008; posted at: <http://plato.stanford.edu/entries/logic-paraconsistent/>).

G. Priest, R. Routley, and J. Norman (eds.), *Paraconsistent Logic: Essays on the Inconsistent.* Philosophia Verlag, Munich, 1989.

V. Propp, *Morfologija skazki,* in *Voprosy poetiki,* 12. Gosudarstvennyi Institut Istorii Iskusstva, Leningrad, 1928. English editions: *Morphology of the Folktale,* ed. by S. Pirkova-Jakobson (transl.: L. Scott), (Indiana University Research Center in Anthropology, Folklore and Linguistics, publication series, 10; Indiana University, Bloomington, Indiana, 1958). Reprinted in: *International Journal of American Linguistics,* Vol. 24, No. 4, Part 3 (Bibliographical and Special Series of the American Folklore Society, 9). New English translation: *Morphology of the Folk-*

tale, 2nd edn., ed. by L. A. Wagner (University of Texas Press, Austin, Texas, 1968.) [American authors are used to refer to the Austin editions.]. Revised Russian edn.: Nauka, Leningrad, 1969; whence French edn., *Morphologie du conte* (collection Poétique; Éditions du Seuil, Paris, 1970).

L. Prussin, 'Judaic threads in the West African tapestry: No more forever?', in *The Art Bulletin,* 88(2), 2006, pp. 328-353.

D. Quaglioni, 'I processi contro gli Ebrei di Trento. 1475. Il procedimento inquisitorio', in: J. C. Maire Vigueur and A. Paravicini Bagliani (eds.), *La parola all'accusato.* Sellerio, Palermo, 1991, pp. 282-294.

A. A. al-Rabe, *Muslim Philosophers' Classifications of the Sciences: al-Kindi, al-Farabi, al-Ghazali, Ibn Khaldun*, Ph.D. Thesis, Harvard University, 1984.

R. Rashed, 'Le commentaire par al-Kindi de *L'Optique* d'Euclide: un traité jusqu'ici inconnu', in *Arabic Sciences and Philosophy: A Historical Journal,* 7(1), 1997, pp. 9-56.

R. Rashed and J. Jolivet (eds.), *Oeuvres philosophiques et scientifiques d'Al-Kindi* (Arabic or Latin text with French translation on facing pages; introduction and commentary in French), in the series Islamic Philosophy, Theology, and Science, 29. Brill, Leiden, 1997. Vol. 1: *L'Optique et la cataoptrique,* ed. by R. Rashed; Vol. 2: *Métaphysique et cosmologie,* ed. by R. Rashed and J. Jolivet.

V. Raskin, 'Semantics of lying', in: R. Crespo, B. Dotson-Smith and H. Schultink (eds.), *Aspects of Language: Studies in Honour of Mario Alinei,* vol. 2: *Theoretical and Applied Semantics.* Rodopi, Amsterdam, 1987, pp. 443-469.

V. Raskin, *Semantics of Lying.* Kluwer, Dordrecht, The Netherlands, and Boston, 1993.

V. Raskin, 'Computer implementation of the General Theory of Verbal Humor', in: J. Hulstijn and A. Nijholt (eds.), *Automatic Interpretation and Generation of Verbal Humor: Proceedings of the 12th Twente Workshop on Language Technology,* Twente, 1996. University of Twente, Enschede, The Netherlands, pp. 9-19.

F. Renda, *La fine del giudaismo siciliano: Ebrei marrani e Inquisizione spagnola prima durante e dopo la cacciata del 1492.* Sellerio, Palermo, 1993.

N. Rescher, *Al-Kindi: An Annotated Bibliography.* University of Pittsburgh Press, Pittsburgh, Pennsylvania, 1964.

D. P. Robertson and R. Cipolla, 'Building architectural models from many views using map constraints', in: A. Heyden, G. Sparr, M. Nielsen and P. Johansen (eds.), *Proceedings of the Seventh European Conference on Computer Vision,* Copenhagen, May 2002. (Lecture Notes in Computer Science, Vol. 2351.) Springer-Verlag, Berlin, Vol. 2, 2002, pp. 155-169.

F. Rosenthal, 'al-Kindī, Abū 'Umar Muḥammad b. Yūsuf al-Tudjībī', in English, in *The Encyclopaedia of Islam, New Edition,* vol. 5, Brill, Leiden, 1986, pp. 121-122.

J. Romm, Review of *Palaephatus: On Unbelievable Tales,* trans. J. Stern (Bolchazy-Carducci Publishers, Wauconda, 1996), in the online periodical *Bryn Mawr Classical Review,* 96.11.4, 1996.

J. L. Russo, *Lorenzo da Ponte: Poet and Adventurer.* Columbia University Press, New York, 1922. Based on a PhD thesis.

J. Schlosser Magnino, *La letteratura artistica,* trans. F. Rossi from *Die Kunstliteratur* (Kunstverlag Anton Schroll & Co., Vienna, 1924), 3rd Italian edition updated by O. Kurz (La Nuova Italia Editrice, Scandicci, Firenze; in the series «Il pensiero storico», 1935; other editions: 1956, 1964; in the series «Paperbacks Classici», vol. 6, 1996).

R. W. Schoenlein, S. Chattopadhyay, H. H. W. Chong, T. E. Glover, P. A. Heimann, C. V. Shank, A. A. Zholents, and M. S. Zolotorev, 'Generation of femtosecond pulses of synchrotron radiation'. *Science,* 287 (2000), p. 2237.

M. Sergot, 'Modelling unreliable and untrustworthy agent behaviour', in: B. Dunin Keplicz, A. Jankowski, A. Skowron and M. Szczuka (eds.), *International Workshop on Monitoring, Security, and Rescue Technique in Multiagent Systems,* Plock, Poland, 7-9 June 2004. Springer-Verlag, Berlin, 2005, pp. 161-177.

G. Shafer, *The Art of Causal Conjecture.* The MIT Press, Cambridge, Massachusetts, 1996.

I. Shahîd and A. F. L. Beeston, 'Kinda', in English, in *The Encyclopaedia of Islam, New Edition,* vol. 5, Brill, Leiden, 1986, pp. 118-120.

A. M. Smith, *Ptolemy and the Foundations of Ancient Mathematical Optics: A Source Based Guided Study.* (Transactions of the American Philosophical Society, vol. 89, pt. 3.) American Philosophical Society, Philadelphia, 1999.

J. Stephen, *The Story of Nuncomar and the Impeachment of Sir Elijah Impey* (2 vols.). Macmillan, London, 1885.

M. Swain, 'Reasons, causes, and knowledge', in *The Journal of Philosophy,* 75(5), 1978, pp. 229-249.

E. Sweetser, 'The definition of *lie*: An examination of the folk theories underlying a semantic prototype', in: D. Holland and N. Quinn (eds.), *Cultural Models in Language and Thought.* University of Chicago Press, Chicago, 1987, pp. 43-66.

A. Tafazzoli, 'King Ferêdûn', at the *Iranian Mythology* website hosted at the Circle of Ancient Iranian Studies (CAIS) of the School of Oriental and African Studies (SOAS) of the University of London. Posted on the Web at the address <http://www.cais-soas.com/CAIS/Mythology/freydun.htm> (last modified: 26 January 2007, when accessed on 1 February 2007).

J. Taylor, *A Multi-Agent Planner for Modelling Dialogue.* Ph.D. thesis, School of Cognitive and Computing Sciences, University of Edinburgh, Edinburgh, Scotland, 1994a.

J. A. Taylor, 'Using hierarchical autoepistemic logic to model beliefs in dialogue', in: J. R. Koza (ed.), *Artificial Life at Stanford 1994.* Stanford Bookstore, Stanford, CA (1994b). Also: Technical report HCRC/RP-60 (November 1994), Human Communication Research Centre, University of Edinburgh, Edinburgh, Scotland.

F. Titi, *Studio di pittura, scoltura et architettura nelle chiese di Roma.* Mancini, Rome, 1674. Reprinted, Rome: Piccini, 1675. Augmented edition: G. Vannucci, 1686. Augmented by F. Posterla, publisher P. Leone, 1708. Edition *Nuovo Studio* (last edition by F. Titi, augmented including palaces), 1721. Last edition: *Descrizione delle pitture, sculture e architetture esposte al pubblico in Roma,* publisher M. Pagliarini, Rome, 1763, reprinted in facsimile with an introduction by F. Prinzi, Multigrafica Editrice, n.p. (but Rome), 1978.

P. Travaglia (ed.), *Magic, Causality and Intentionality: The Doctrine of Rays in al-Kindi.* Micrologus Library, 3. Edizioni del Galluzzo, Florence, for the Società internazionale per lo studio del medioevo latino (SISMEL); distributed by Brepols, Turnhout (Belgium), 1999.

G. Treasure, *Who's Who in Early Hanoverian Britain.* (Who's Who in British History, 6.) Shepeard-Walwyn, London, 1992.

G. Troupeau, 'al-Kindī, 'Abd al-Masīḥ ibn Iṣḥāq', in English, in *The Encyclopaedia of Islam, New Edition,* vol. 5, Brill, Leiden, 1986, pp. 120-121.

H.-J. Uther, *The Types of International Folktales: A Classification and Bibliography. Based on the System of Antti Aarne and Stith Thompson. Part 3: Appendices* (Folklore Fellows Communications, 286) Suomalainen Tiedeakatemia = Academia Scientiarum Fennica, Helsinki, 2004.

G. Vasi, *Itinerario istruttivo di Roma diviso in 8 giornate.* Rome, 1763. Reprinted until 1820.

A. Velupillai, 'The Vision of Civan in Tamil Caivam', at the *Tamil Nation* website (posted at <http://www.tamilnation.org/sathyam/east/saivaism/vision.htm>).

VENIVA, 'Retimo e la rappresentazione della città' (posted on the Web at the webpage whose address is <http://www.msc.gr/veniva/it/c2/doc5.htm>), part 5 of 6 in: 'L'immagine e la città: La guerra di Candia', at the website *Viaggio virtuale tra le fonti storiche veneziane – Rotta: Venezia e il Levante (sec. XV - sec XVIII).* The VENIVA consortium, © 1996 (accessed in Jan. 2007).

A. Verhagen, *Constructions of Intersubjectivity: Discourse, Syntax, and Cognition.* Oxford University Press, Oxford, 2005. Reviewed in Fritz (2006).

V. R. Walker, 'Complexity, transparency, and the warranted use of formal systems in legal factfinding', in: A. A. Martino and E. Nissan (eds.), *Formal Approaches to Legal Evidence*, special issue, in *Artificial Intelligence and Law,* 9(2/3), 2001, pp. 189-197.

K. Walton, 'Fearing fictions', in *The Journal of Philosophy,* 75 1978, pp. 5-27.

K. Walton, *Mimesis as Make-believe.* Harvard University Press, Cambridge, Massachusetts, 1990.

C. Wilkinson: *Forensic Facial Reconstruction.* Cambridge University Press, Cambridge, U.K., 2004.

J. H. Woods, *The Logic of Fiction.* Mouton, The Hague, The Netherlands, 1974.

J. Woods and P. Alward, 'The logic of fiction', in *Handbook of Philosophical Logic,* 2nd edition, Vol. 11, ed. D. Gabbay and F. Guenthner. Kluwer, Dordrecht, The Netherlands, in press.

M. E. Yapp, Hilmand (in English), in *The Encyclopaedia of Islam, New Edition,* vol. 3. Brill, Leiden, & Luzac, London, 1979, p. 392.

E. Yassif [עלי יסיף] (ed.), *The Tales of Ben Sira in the Middle Ages: A Critical Text and Literary Studies*, in Hebrew: סיפורי בן סירא בימי הביניים The Magnes Press, Jerusalem, 1984.

E. Yassif, *The Hebrew Folktale.* Hebrew original: The Bialik Institute, Jerusalem, and Ben-Gurion University Press, Beer-Sheva, Israel, 1994. English translation by J. S. Teitelbaum. Indiana University Press, Bloomington, Indiana, 1999.

Endnotes

1 The philosopher Ya'qūb ibn Iṣḥāq al-Kindī [Jolivet and Rashed 1986] is not to be mistaken for the Arabic-language Christian theologian known as 'Abd al-Masīḥ ibn Iṣḥāq al-Kindī. The latter is 'the name given to the author of a defence of Christianity presented in the forms of a letter written in response to that of a Muslim friend, [...] who invited his correspondent to embrace Islam', and variously dated to the ninth or early tenth century [Trupeau 1986]. Another al-Kindī had as his scholarly concern the history of Egypt (897-961); see Rosenthal [1986].

The philosopher Ya'qūb ibn Iṣḥāq al-Kindī – whose full name was Abū-Yūsuf Ya'qūb ibn Iṣḥāq ibn as-Ṣabbāḥ ibn 'Umrān ibn Isma'īl al-Kindī – died in 873. A descendant of the Kinda Arabian tribe (from Yemen), he was born in Kufa (in Iraq) perhaps in 801. He was educated in Kufa and then Baghdad. He was appointed to oversee the translation of Greek texts of learning. This was a policy pursued by the Abbasid caliphs of Baghdad, whereas the previous dynasty, the Umayyad caliphs of Damascus – Bruno Chiesa pointed out [Chiesa 1989, pp. 15-16] – had been reluctant to make official concessions to Greek culture, precisely because the previous official language of Syria (under Byzantine rule) had been Greek.

As a mathematician, Ya'qūb ibn Iṣḥāq al-Kindī was among the first to introduce the use of Indian numerals – he contributed to their spread, through his treatise *On the Use of the Indian Numerals (Kitāb fī Isti'māal al-'Adād al-Hindī)* – and moreover he was active in cryptography. 'Al-Kindi wrote many works on arithmetic which included manuscripts on Indian numbers, the harmony of numbers, lines and multiplication with numbers, relative quantities, measuring proportion and time, and numerical procedures and cancellation. He also wrote on space and time, both of which he believed were finite, 'proving' his assertion with a paradox of the infinite. [Garro (1978)] gives al-Kindi's 'proof' that the existence of an actual infinite body or magnitude leads to a contradiction [whereas] Garro [1994] formulates the informal axiomatics of al-Kindi's paradox of the infinite in modern terms and discusses the paradox both from a mathematical and philosophical point of view. In geometry al-Kindi wrote, among other works, on the theory of parallels. He gave a lemma investigating the possibility of exhibiting pairs of lines in the plane which are simultaneously non-parallel and non-intersecting. Also related to geometry [were] the two works he wrote on optics, although he followed the usual practice of the time and confused the theory of light and the theory of vision' [O'Connor and Robertson 1999].

As a physician, Ya'qūb ibn Iṣḥāq al-Kindī introduced a scale of drug potency. His medical writings were discussed by Demirhan Erdemir [2002], who is concerned with al-Kindi's *al-Aqrābādhīn (The Dispensatory)*, an English translation of which exists [Levey 1966]. It is not to be mistaken for a pharmaceutical work by

Sābūr ibn Sahl ibn Sābūr (d. 869), *al-Aqrābādhīn,* of which the short version *(Dispensatorium parvum)* is preserved and was edited by Kahl [1994]. When it comes to chemistry, it is important to observe that al-Kindi was an opponent of alchemy, and refuted the transmutation of metals.
As a philosopher, al-Kindi dealt with subjects important to theologians, but in philosophy he was overshadowed by al-Farabi, and this was among the likely reasons for some of his philosophical works being lost, along with his downfall under the caliphs al-Wathīq (reg. 842-847) and al-Mutawákkil (reg. 847-861). Al-Kindi's downfall was, according to some, caused by scholars' rivalry, rather than by al-Mutawákkil's fanatism, and his overthrowing the Mu'tazilite creed (within Sunni Islam) as state religion. At one point, al-Kindi was beaten, and for a while his library was assigned to rival scholars. At any rate, al-Kindi did not recover his status. Yet, al-Kindi's lasting epithet has been 'the Arab philosopher' *par excellence.* It also appears in the title of books by Gustav Flügel [1802-1870] [Flügel 1857] and by George Atiye [1994]. For recording for al-Kindi this epithet, credit is given to the historian Ibn al-Naḍīm (d. 955). Actually, the epithet is explained by al-Kindi's having been the only notable philosopher whose genealogy went back to the Arabian peninsula, rather than to acculturated people. The Kinda tribal group was Arab, unlike other groups also from Southern Arabia. In the centuries before the rise of Islam it spread all over Arabia, and was known as Kindat al-Mulūk, 'Royal Kinda'. The name *Kinda* of the alleged ancestor was a nickname for *Thawr* ('Bull'). See Shahîd and Beeston [1986].
Al-Kindi's interpretation of Aristotle was Neo-Platonic. Guerrero [1992] contrasts how al-Kindi and al-Farabi handled Aristotle's *De anima.* Jean Jolivet [1971] is concerned with the intellect according to al-Kindi. In western Europe, Ya'qūb ibn Iṣḥāq al-Kindī was known by the Latin name *Alkindus.* Some of his works only survived in a Latin translation by Gerard of Cremona. The scientist Gerolamo Cardano (1501-1575) enumerated him among the twelve most prominent scholars of the Middle Ages.

2 The three quotations are from the publisher's blurb of Pinella Travaglia [1999]; e.g., in Brepols [2003, p. 17]. They are slightly edited, especially the punctuation. The typo 'casual' for 'causal' is a frequent one, and I have come across it even in the abstract of a highly mathematical article on causality and Bayesianism, for a journal special issue I was guest-editing. Of course the distinction between 'casual' and 'causal' is paramount.

3 I.e., al-Kindi's physical approach in *On Rays,* as opposed to a different view which he also entertained. The Wikipedia entry on al-Kindi (posted at the site <http://en.wikipedia.org/wiki/Al-Kindi>) points this out: 'However, he is ambiguous when it comes to the actual process by which the heavenly bodies affect the material world. One theory he posits in his works is from Aristotle, who conceived that the movement of these bodies causes friction in the sub-lunar

region, which stirs up the primary elements of earth, fire, air and water, and these combine to produce everything in the material world. An alternative view found in his treatise *On Rays* is that the planets exercise their influence in straight lines. In each of these, he presents two fundamentally different views of physical interaction; action by contact and action at a distance. This dichotomy is duplicated in his writings on optics'. A. Mark Smith's book [1999] is on Ptolemy's optics. Cultural meanings associated with light, as well as explanations given for its nature, throughout the history of Western civilization, are discussed by Park [1997]. Al-Kindi concerned himself with optics and human vision, and his theory of vision is the point of departure for Lindberg [1976]. Roshdi Rashed described in a paper [Rashid 1997] al-Kindi's commentary to Euclides' *Optics,* and also edited a French edition of al-Kindi's optics, and a companion volume he edited with Jean Jolivet is about cosmology and metaphysics [Rashed and Jolivet 1997]. Al-Kindi also engaged in weather forecasting [Bos and Burnett 2000]. This is apt for our point: if you don't have a theory, what the weather will be is entirely random, instead of just partly so.

Sylvain Matton [1976] has a chapter on 'al-Kindi: *Des rayons;* ou, *Théorie des arts magiques*', in a volume on Arab magic which also contains passages from Ibn Khaldūn and Ibn Waḥshiyya.

Nicholas Rescher published [1964] an annotated bibliography of 55 pages on al-Kindi. A Harvard dissertation by Ahmad al-Rabe [1984] contrasts the classification of the sciences according to the medieval islamic scholars al-Kindi, al-Farabi, al-Ghazali, and Ibn Khaldun.

4 Negrotti [2000, p. 295, Table 1; 2002, p. 69] distinguishes between (A) the *concrete artificial,* consisting of material devices or processes, and (B) the *abstract artificial,* consisting of informational devices or processes. He also distinguishes between (a) the *analytical artificial,* which reproduces structures, and (b) the *aesthetic artificial,* which reproduces appearance. Robots and virtual reality (when the latter interacts with the real world) belong in (Aa). Also a prothesis replacing a body organ does, and so do pharmaceuticals. Category (Ba) includes artificial intelligence, as well as subsymbolic reasoning techniques. Examples of (Ab) include, e.g., sculpture; note that the purpose is not necessarily properly artistic, as arguably also facial reconstructions (either concrete, or as a computationally obtained 3D image), which reproduce a human individual's face based on the concrete evidence of skeletal remains (this being a rather controversial technique from forensic pathology: see Wilkinson [2004]) also belong in category (Ab), and their purpose is eminently practical: the identification of a given anonymous human body (even though on occasion the technique has been applied in order to show how some historical character may have looked like when alive).

Examples of category (Bb) include: drawing, maps, diagrams obtained by simulations in a space of parameters, as well as virtual reality when it is merely descriptive. Arguably, both (Ba) and (Bb) are relevant categories for three-dimensional (3D) animations of the human body that incorporate semantics and are based on anatomically realistic models, such as, e.g., in Aubel and Thalmann [2000] and Gutierrez et al. [2005]. As to the reduction of 3D real scenes to maps or photographs in two dimensions, consider that the state of the art of machine vision enables to some extent the reconstruction of architectural scenes from photos (even uncalibrated) and maps [Infantino et al. 2000, Robertson and Cipolla 2002]. Category (Aa) of Negrotti's classification also includes tissue engineering, and the use it makes of polymers. 'Tissue engineering is a discipline that probes the structure-function relationship of normal and diseased tissue in an effort to restore, maintain or improve the function of living tissue' [Baldwin and Saltzman 1996, p. 177].

Nissan [2000] discussed meat surrogates or more generally, biomimetics – this belongs in category (Aa), and the artificial is both by structure and by process [but for an overview of biomimetism in robotics, see Cassinis, Morelli and Nissan 2007] – as well as surrogates of writing (a manual algorithm with concrete data structures) under given culture-bound constraints. Then the artificial is by process; category (Ba) is relevant. The observation level is formal, syntactic. Moreover, Nissan [2000, Sec. 5] discussed some classes of theatrical representations: in a genre of Burmese dance, the movements of human dancers are such as to imitate a puppet-show; several actors who perform in Karel Čapek's drama *R.U.R.: Rossum's Universal Robots* [Čapek 1921], play in the role of robots (it was that play that introduced the term *robot*). The artificial is by appearance. The observation level, in the case of the Burmese dance, is kinetic: dis/continuity in body movements [this is an observation level which is relevant also for *Survivor,* a walking chair from robotic art that symbolises persons who survived landmine blasts as amputees: see Cassinis et al. 2007]. Paradoxically, the Burmese dancers or the actors in *R.U.R.,* even though they are human, enact a reproduction of a kind which Negrotti calls *technoids*, i.e., technological devices. Yet, those devices in turn (puppets or robots) are *naturoids,* because they reproduce features from natural exemplars, namely, of humans.

5 There had already been plague epidemics in Milan in 1398-1401, 1502, and 1524-1528. There were ten such epidemics all in all between 1378 and 1629-1623 in Milan. It was in 1576 that the official belief in the *untori,* i.e., the deliberate propagators of the illness, first appeared in Milan.

6 The dynamics of the coalesced mind, as though, of a crowd, which itself may become, e.g., an angry mob, has been formalized computationally by Andrew Adamatzky [2005]. He has modelled mathematically collective emotions and how these affect rationality. At the simpler level of his treatment, he represents the dy-

namics of affects and beliefs respectively by means of *affectons* and *doxatons*. '*Affecton* is a finite automaton whose internal states, input and output states, are *happiness, anger, fear, sadness, confusion* and *anxiety*' [*ibid.,* p. 227]. '*Doxaton* is a finite automaton whose internal states, input and output states, are *knowledge, doubt, delusion, misbelief* and *ignorance*' [*ibid.,* p. 228], where *delusion* is defined as 'an erroneous belief that cannot be justified' [*ibid.*], '*Misbelief* is a wrong belief' [*ibid.*], '*Knowledge* is a justified belief' [*ibid.*], and '*Doubt* is a state of neither belief nor *misbelief* about a locally justifiable fact' [*ibid.*]. '*Doxastic world* is an element of a set of all possible sets of *doxastic states*' [*ibid.*], where '*Doxastic state* is an agent or automaton state derived from belief, which includes' the five kinds enumerated under *doxaton* [*ibid.*]. Domination relations among doxastic states lead to a state being extinguished [*ibid.,* p. 149], under an assumption of equiprobability of the five doxastic states at the beginning of development. Thus: 'In these conditions, misbelief, doubt and ignorance are absorbed by other doxastic states if knowledge dominates misbelief and doubt, misbelief dominates doubt and delusion dominates ignorance. To exterminate belief and doubt, we must apply such combinations of norms that knowledge dominates doubt and misbelief, misbelief is dominated by doubt and ignorance dominates delusion. The knowledge becomes extinct if it is dominated by doubt and misbelief. The delusion vanishes if it is dominated by ignorance and knowledge is dominated by doubt and misbelief. Misbelief can not be found in the development only if it is dominated by knowledge and doubt. The situation with doubt is intriguing. Doubt becomes extinct if it is dominated by knowledge, dominates misbelief and knowledge dominates misbelief' [*ibid.,* pp. 148-149].

7 Alessandro Manzoni's novel *I Promessi Sposi* has made the concept of the *untori* – i.e., 'smearers' who during the plagues of the 17th century were supposed to deliberately spread the contagion – familiar to Italian school-leavers, but in that context the *untori* were not specifically associated with an ethnic or denominational minority (perhaps because in the Spanish-dominated Duchy of Milan in Lombardy, the small Jewish population had recently been expelled, a major role having been played in that by St Carlo Borromeo, the archbishop of Milan).

8 Canon Giovanni Di Giovanni, who in other respects was an historian of some merit, was Vice-Inquisitor General of the Kingdom of Sicily: see Cusumano [2004, Sec. 3, pp. 97-104], who points out how Di Giovanni made a spurious claim that the alleged event in Messina eventually was a major factor in determining the expulsion of the Jews from Sicily in 1492. Actually the expulsion was decided upon by the King of Aragon, in concomitance with the expulsion of the Jews from Spain and Sardinia, and in the main unwelcome to the local authorities in Sicily: in fact, these fully realized that a precedent had been set, as by taking such a step without seeking the agreement of the islanders, Ferdinand of Aragon dispelled the notion that the island retained the autonomy they had trusted or

hoped it did, notwithstanding exactly one century of the island belonging to the Crown of Aragon.

9 The decree of 1740 was followed in 1746, because of insistent pressures, by a decree of expulsion [Giura 1984]. Cusumano remarks that in the 19th century, instead, Romantic historiography, represented by La Lumia [1883], preferred to claim that in 1492, the Sicilian populace in tears bade farewell to the expelled Jewish population. See now Francesco Renda's book [Renda 1993].

10 The circumstances of the plague-time libel instantiate a pattern of a group, delimited by illness (lepers) or denomination and ethnicity (Jews), being blamed as either the cause or the beneficiaries of a physical catastrophe. Something similar also happened for a political catastrophe (such as defeat in World War One). Another instance is the Korean immigrant community in the immediate aftermath of the Kanto Quake, i.e., the Tokyo and Yokohama earthquake and fire of 1 September 1923 (Kanto is the plain of Tokyo and Yokohama). Well-poisoning accusations, familiar from medieval Europe, were levelled in that case, too (and this led to a massacre of Koreans by rampaging crowds), but it is an example such that it can be seen that that particular accusation was only one concomitant option along with different ones, also actualized (e.g., accusing Koreans of looting after the earthquake).

A good case in point is the trial of health workers from the children's hospital in Benghazi, Libya. In February 1999, 23 Bulgarian health workers from that Benghazi hospital had been detained in connection with an investigation into an increased incidence of AIDS at that hospital. This was when the Bulgarian diplomatic authorities became aware of their nationals' detention (during the previous autumn, there had been some such detentions, followed by release). Fifteen of the 23 Bulgarians detained in February were released, but six (five nurses and one doctor) were held in custody. Two nurses made confessions, later retracted.

Important allegations were also made, since, about the mode of interrogation. Rumours about nosocomial infection (i.e., such that it was being acquired at the hospital) had caused panic among local residents in summer 1998. A Libyan magazine report found that at the Benghazi pediatric hospital, there was a shortage of consumables, disposable instruments were repeatedly used, no AIDS tests were conducted, and basic rules of hygiene were not observed. This (by a reasonable conjecture) led to a spread of the disease among children outpatients or patients. In the autumn of 1998, it was alleged for the first time that the children had been infected *deliberately* as part of a conspiracy against the Libyan state. From tests on Benghazi infected children conducted by collaboration with hospitals in Europe, it resulted that infection included HIV and kinds of hepatitis, as well as various combinations of these, and that the infection was nosocomial indeed.

In February 2000, six Bulgarians, nine Libyan nationals and a Palestinian doctor were charged with *deliberately* infecting 393 Libyan children with HIV, and the

Bulgarians were also charged with the perpetration of acts incompatible with Libyan norms and traditions. (This suggests that the foreignness of the foreign workers played a role in considering their practice or malpractice as being malevolent). Six death sentences, for five Bulgarian nurses and a Palestinian doctor, were confirmed on 19 December 2006, and the defendants were expected to appeal to Libya's High Judicial Council. The condemnation drew a strong protest from the Bulgarian government and president, stating that a guilty verdict could not be accepted, and that it was a blow to the Bulgarian nation. On 17 July 2007, Libya's High Judicial Council commuted into life imprisonment the death sentences of the five Bulgarian nurses and the Palestinian doctor (they had been in prison since 1999), whereas the families of infected children were each being indemnified with a conspicuous sum by an international fund in which Libya participated along with countries from the European Union. It was pointed out that the hundreds of millions, or, it was also reported, billions of pounds this involved happened to be an amount similar to an amount that Libya was claiming Bulgaria owed her. Finally, on 23 July 2007, the six defendants were sent to Bulgaria. France's President Sarkozy was credited for negotiating and obtaining their release.

11 In the entry 'Juif' in Diderot and D'Alembert's *Encyclopédie* (1751-1772), an entry authored by Louis de Jaucourt, mentions of the last two accusations (which it was far from crediting) are juxtaposed to each other: 'on les accusoit de magie, de sacrifier des enfans [sic], d'empoisonner les fontaines' [Cusumano 2004, p. 81].

12 After 1840, Manzoni refrained from further writing fiction. From then on, he focused on essays in historiography, aesthetics, and language policies. By then, he was finding unfeasible his early goal of writing novels while not prevaricating on historical truth (see his essay of 1845, *Del romanzo e in genere de' componimenti misti di storia e d'invenzione*).

13 In *Storia della colonna infame,* Manzoni reconstructed the circumstances of the execution of a barber who had been charged with being an *untore* and spreading the plague; the 'infamous column' being the monument which commemorated his conviction and execution. Whereas the monument expressed the view that the man was guilty, unsurprisingly to Manzoni he was innocent.

14 After the publication of the *Storia della colonna infame,* in which Manzoni exposed the falsehood of the myth of the *untori,* Samuel David Luzzatto, the director of the Collegio Rabbinico in Padua, turned to him in a letter that reportedly remained unanswered, asking him to similarly expose the falsehood of the accusation of ritual murder as in the 1475 case from Trento. It is also alleged that Manzoni was approached over this matter and declined, stating that he didn't know whether the Jews accused in Trento were guilty or innocent. Remarkably, he didn't want to probe further in order to find out. On the Trento libel of 1475, see, e.g., Hsia [1992], Esposito and Quaglioni [1990], Quaglioni [1991].

15 Apart from his clerical leanings (in his youth he rather was a freethinker), Manzoni may have been reluctant to challenge what by his times was official doctrine, as stated in Benedict XIV's papal bull of 1755, *Beatus Andreas,* about the alleged ritual murder of the child Andreas of Brixen, i.e., Bressanone in South Tirol, in 1462, a few years before the libel of 1475 concerning the child Simonino in Trento: an honest envoy of the Pope right after the Trento trials – the apostolic commissar Battista de' Giudici, a dominican friar, bishop of Ventimiglia, a Thomist theologian by formation, and an adversary of the humanists Bruni and Platina – had found them to be grossly irregular, and the accusation itself preposterous, but a cover-up prevailed. On other grounds, de' Giudici had preached against the Jews, so his defence (as well as earlier papal bulls prohibiting that kind of accusations) should have been credible. The Trento trial proceedings themselves explicitly endorse a *pia fraus,* 'a pious fraud', for the sake of obtaining a conviction. The itinerant preacher Bernardino of Feltre had 'prophesied' that before Easter, the crimes of the Jews would be uncovered. Hessen-born Johannes Hinderbach, the Prince-Bishop of Trento, was determined to obtain a conviction and a local saint (the child Simonino), and actually as a result of his propaganda he acquired such influence in Italy that would have been unobtainable otherwise. At the time of the trials, the Doge of Venice expressed, in a letter, concern and pointed out how different the attitudes of Venice were. Yet, a few years later on, the populace in the Republic of Venice was swayed enough by the example and active propaganda of Trento, for copycat accusations to be made, and for convictions to be obtained, in Marostica and in Portobuffolè, with executions being carried out, or for a Hungarian Jew to be thrown into a Venice canal by an angry mob. Trendy humanists (like de' Giudici's adversary, Platina) were happy to join in the new climate of demonization of a minority.

16 Clement XIII succeeded Benedict XIV in 1758. Cardinal Ganganelli presented a report, dated 24 December 1759, denouncing ritual murder accusations against Jews as false, but did not denounce also the cases of Andreas of Brixen and of Simonino of Trento, because of the bull *Beatus Andreas* (see notes 15 and 17). The Church approving of a local cultus did not entail that the martyrdom was true.

17 As the historian Dubnov, a classic, pointed out, nowhere there had been as many blood libel trials as in Poland in the short period from the 1730s and the 1750s. The blood libel of 1710 had been believed by King Augustus II of Saxony-Poland. Yet, blood libels had been disavowed by various Popes in previous ages. In 1760 this prohibition was reconfirmed (as a response to the spate of trials in Poland).

Cardinal Ganganelli, the future Pope Clement XIV (b. 1705, r. 1769, d. 1774), submitted in January 1760 a relation to Pope Clement XIII, who approved it. Ganganelli remarked about the similarity of Roman-age blood libels against members of the early Church, and the current ones in Poland. In his relation, he proposed to reaffirm the previous papal disavowals of anti-Jewish blood libels, disavowals promulgated as early as Innocent IV (1243-1254). However, Cardinal Ganganelli preferred not to challenge the by then entrenched cult of Simonino of Trento (it was only abolished by the Church in 1965). It would have been unnecessary in the given circumstances (with the ongoing events in Poland) to take on a traditional celebration in the liturgical calendar of the Prince-Bishop of Trento, especially as just a few years before, Benedict XIV had enshrined in doctrine cults such as the child Simonino's in his bull of 1755, *Beatus Andreas* (which may well have encouraged the spate of blood libels ongoing at the time in Poland). Rather, Ganganelli's report claimed that the Trento case must be confined to an individual event of assassination out of hatred ('in odium fidei' was the expression used by the 1755 bull).

Actually, a bull of Pope Innocent IV, from 5 July 1247, disavowed unequivocally the blood libel, and identified greed as being the motivation behind those making it (as it would enable them to seize the assets of an accused local Jewish community). The outstanding exception to this policy of the Popes reining in blood libels on doctrinary grounds, was Innocent III, in office in 1199-1216 (in 1209 he declared the crusade against the Albigenses). He did believe in blood libels. Asking whether this determined his anti-Jewish attitudes as reflected by his policies, or the other way around, is like asking what came first: the chicken or the egg.

18 Yet, Salieri was benevolent with Mozart, and malevolent with Da Ponte. Shared Italianness did not result in Salieri and Da Ponte being allies, *pace* Mozart's prejudice. On Lorenzo Da Ponte, see Holden [2006], Hodges [1985], FitzLyon [1955], Russo [1922], Marchesan [1900], Lanapoppi [1992].

19 Carlo Emilio Gadda (Milan, 1893 - Rome, 1973) was Italy's greatest modernist writer. Because of how he uses language, his works are very difficult to translate.

20 The story in this novel by Gadda takes place in 1927 (when both he and the protagonist were 35).

21 The aim of Swain [*ibid.*, p. 229] is 'providing a combined causal/defeasibility analysis of knowledge'. In his view, 'a person has knowledge *h* only if that person justifiably believes that *h* on the basis of reasons and only if the having those reasons results from a nondefective causal ancestry. A causal ancestry is nondefective if and only if the belief is indefeasibly justified' [*ibid.*].

22 In the tenth century, the Hebrew-language *Book of Josippon* aetiologized the name of Sorrento/Surriento (near Naples) after a character, Sur, supposedly escaping a biblical narrative. The modern *The Antiquities of Kittim* (within the literary corpus used in GALLURA, and authored by myself in late-antique or medieval strata of Hebrew in a pseudo-midrashic style, i.e., emulating traditional Jewish homiletics) includes hundreds of aetiological tales, in which an explanation is concocted for Italian placenames, usually by interpreting the names in Hebrew. Most such tales are concise; here and there, a longer tale occurs.

For example, a story about the biblical giant Og (legendarily reworked in the midrashic literature) was embedded in the discussion of the Sardinian town of Perdasdefogu (actually called after 'stones of fire': coal). The placename is interpreted as Hebrew *Pe'ér da'*: *sde 'af 'Óg hu,* i.e., 'Know [our] glory: it is the field of Og's nose'; a suitable story is told. Italy has about 8,000 municipal councils; Italy's toponyms by the hundreds were mythopoietically aetiologized in our corpus, region by region, and province by province. The names of all sixty municipal councils of the new province of Lodi, in southern Lombardy, lent themselves to mock-explanation in satisfactory short tales.
This was done by a human author (myself). What can be automated, of such writing strategies and tactics? Sorting into a typology of devices is called for. The GALLURA project gives folk-etymological aetiological devices a 'respectability' which in given historical-cultural contexts they deserve. Theories of explanation in general would benefit from explicit incorporation of genre-bound poetic conventions and non-bona-fide (e.g., mock-) aetiological intentionality.

23 Sergot [2005] is on logic-based modelling of unreliable and untrustworthy agent behaviour. Lying and deception from the viewpoint of forensic psychology are the theme of de Cataldo Neuburger and Gulotta [1996]; cf. Castelfranchi and Poggi [1998], whose perspective is that of cognitive science. Eve Sweetser [1987] is concerned with the definition and the semantic prototype of 'lie'. Also Raskin [1987, 1993] is concerned with the semantics of lying. For the computer modelling of *trust and deception* in a society of agents, see, e.g., Castelfranchi and Tan [2002]. Argumentation in deceptive communication is treated in Carofiglio et al. [2001], Carofiglio and de Rosis [2001]. Floriana Grasso [2002a] discusses fairness and deception in rhetorical dialogues; Grasso [2002b] is more generally concerned with computational rhetoric. In *cooperative dialogues,* i.e., such dialogues that none of the participants is committed to any form of deception, three levels of nesting of beliefs are sufficient. This is not enough when it comes to modelling some situations involving deception, and deeply-nested belief levels are required, as argued by Jasper Taylor [1994a, 1994b]. Modelling and evaluating trust in a public key infrastructure, matter in computer security [Basden et al. 2001, Chadwick and Basden 2001, Ball et al. 2003, Chadwick et al. 1998].

24 The island of Crete was ruled by Venice from the early 13th century, to its fall to the Ottomans in 1669. It used to be referred to as Candia, which also was the name of its chief town, now called Iraklion (Ηρακλειον). Another important and fortified city was Canea (Chania, Χανία, in Greek).

25 In Venetian sources, the town is also referred to as *Rettimo,* but as Venetians degeminate (i.e., double consonants are pronounced as simple), the spelling *Rettimo* must be due to hypercorrection (and perhaps also to patterning the spelling, by attraction or by a portmanteau process, after the Italian adjective *settimo* 'seventh').

26 Rethymni is on the northern coast of Crete, like Iraklion and Chania. Rethymni is half-way between Iraklion (which is on the central longitude of Crete), and the town of Chania in the western part of the island.

27 To the Venetian *provveditore* Luca Michiel, Retimo was miserable and desolated.

28 The map is at the Biblioteca Museo Civico Correr, Venice, Portulan no. 4 (posted at the webpage: <http://www.msc.gr/veniva/panel/p211.htm>).

29 Raskin [1996] and Attardo in their works have shown how, basically, it is amenability to competing scripts (loosely meant), that underlies humour. Victor Raskin's 'Script-based Semantic Theory of Humor' (SSTH) evolved into Attardo and Raskin's 'General Theory of Verbal Humor' (GTVH). Attardo [1996], discussing non-joke humour, mentioned (p. 95) the existence of an 'expansionist approach' (to which neither he, nor I would commit ourselves), according to which, non-joke humour could be treated as complex jokes. Yet, 'the broad theory to be developed will [possibly] need to incorporate some technical apparatus which was unnecessary in the SSTH and GTVH' (p. 87). In SSTH, two claims on verbal jokes were combined: 'such a text is compatible in part or in full with two different scripts' ('script' being used loosely), and 'these two scripts are opposite in a specially defined sense' [Raskin 1996, p. 9]. Moreover (p. 11), in GTVH 'the text of the joke and the script opposition were redefined as two of six knowledge resources (KR's) constituting a joke, Language (LA) and Script Opposition (SO), respectively'; 'four more knowledge resources were introduced, namely, Narrative Strategy (NS), Target (TA), Situation (SI), and Logical Mechanism (LM)'. The latter does not appear in all jokes. Cf. Attardo and Raskin [1991].

30 A jocular riddle on top of that primary riddle could be: 'Wrong! You really don't know Betsie and Leo. They would never ever be second to anybody!'. This is an example of *meta-riddle*. This concept has been discussed by Andrea Fassò [1993]. Concerning his notions of 'metaindovinello' (meta-riddle) and 'metabarzelletta' (meta-joke), see Sec. 8 ('Metaenigmistica') in that paper [*ibid.*, pp. 45-46], as well as endnote 42 on p. 50 in the same article. This should not be mistaken for what elsewhere, in the literature of humour research, has been referred to as *meta-jokes.* Attardo [2006, p. 587] remarks that 'all speakers within the cultural community of English speakers [...] agree that there are jokes that function on the basis of violating the expectation that a given text is a joke (they are known as meta-jokes; an example are shaggy dog jokes). So, any axiomatic account of jokes will have to include some form of theory of types, or similar mechanisms. To put it differently, whatever rules one finds for jokes, there will always be other jokes that violate them. I [Salvatore Attardo] described some of these jokes in Attardo [1988]'.

31 Hilmend or Helmand is the 'name of a river (the Etymandrus of Arrian, the Erymanthus of Polybius, the Haētūmat of the Avesta, the Hīdhmand of the *Ḥudūd al-'ālam*) which, with its five great tributaries (Khūd Rūd, Tirīn, Arghandāb, Tarnak, Arghasān) drains all south-west Afghānistān [...]' [Yapp 1979].

32 Philosophical and computational logicians have been concerned with the modelling of inconsistent beliefs. A quick introduction to paraconsistency and paraconsistent logics, Priest and Tanaka [2000, revised in 2007], is the entry for the concept in the online *Stanford Encyclopedia of Philosophy.* Graham Priest's entry [2002] in the new edition of the *Handbook of Philosophical Logic* is an obvious point of entry into the subject. Among Priest's relevant publications, there is the book Priest [1987]. Cf. Priest et al. [1989].

33 Such homiletics is also found among the Fathers of the Church, and it has been noticed that apparently by the fourth century they were still tapping the resources of Jewish homiletical biblical exegesis, whereas later on, Christian homiletics relied on what was found in the patristic literature.

34 The story of the three youths in the furnace, from ch. 3 of *Daniel,* also had wide currency in Christian traditions. Shown in Fig. 12 is a medieval wall painting (with a detail) from a Nubian church, now on display at the Khartoum Museum. It depicts the story of the three youths thrown into the furnace, with an angel.

35 Apart from early rabbinic culture, also in Hellenistic Greek culture – among the Alexandrian scholars – there used to be a tradition of etymological wordplay. Philip Hardie [1999] reviews a book by James J. O'Hara, who relates this tradition to wordplay in Vergil, thus, in a Latin author from Augustan Rome.

36 The passage is in Aramaic, which is close enough to Hebrew for the verbs for 'to bark' and 'to blow up' to occur with respectively identical lexical roots in both languages. As to the Aramaic verb *netsar* for 'to chirr', 'to chirp', as applied to

Fig. 12. The three youths in the furnace, from a Nubian wall painting (see note 34).

the cricket (Tannaitic Hebrew *tsartsúr,* Modern Hebrew *tsratsár*), in Tannaitic Hebrew it occurs once, and then as denoting making sounds as associated with another animal kind, pigs: the participle *notsr*[*in*] in that sense occurs in the *Mekhilta de-Rabbi Ishmael* – but then not in all versions: it does occur in the Lauterbach edition – within a sentence addressing a character used to eat pork: 'Pigs are grunting from between your teeth' [i.e., it is known you eat them] [Moreshet 1980, s.v. *nṣr*$_2$]. On the *Mekhilta de-Rabbi Ishmael,* see [Kadushin 1969, Neusner 1988a, 1988b, 1990, Lauterbach 1933-1935, Horovitz and Rabin 1960]. Moreshet cites Saul Liebermann, who before interpreting נוצר as being an abbreviation for the plural participle נוצרין (the identification of this verb being plausible on etymological grounds), had initially expressed the opinion that an emendation was required, so that one would read there the participle of the verb for 'to bray' (as applied to asses), which is written נוער (thus differing by just one letter, which is shaped similarly). There are tacit matches involved, too, in the homily about Nebuchadnezzar. The initial part of the personal name *Nevukhadnétsar* matches the Hebrew phrase *nafúaḥ ke-khád* 'blown-up / distended / pot-bellied like a pitcher', as well as the mixed Hebrew and Aramaic phrase (but you could make it fully Aramaic) *navóaḥ ke-'ḥad* [*kalba*]', i.e., 'to bark like [Aramaic:] "one dog".' In fact, *ḥad kalba* is how the text of the homily is wording its equating Nebuchadnezzar to a dog. Incidentally, rabbinic texts, especially in homiletics, tended to acronymize *Nebuchad-Nezzar* into *N.N.* – נ"נ – which by chance in European cultures is the acronym for *Nullius Nominis,* 'nameless', 'of no name' (or, according to the usual interpretation in Italian, *non noto,* 'not known'), which is a bureaucratic indication of unknown paternity.

37 The Latin title *Pseudo Sirach* was coined by M. Steinschneider in the mid 19th century, in order to differentiate the medieval tales of Ben Sira (Sirach) from *Sirach* from the Bible Apocrypha (which is not part of the Hebrew Bible). This is because of the pseudepigraphical ascription of these medieval tales to the biblical Ben Sira. Yassif [1984, pp. 4-6] discusses how this work was named.

38 Eli Yassif [1984] dated it to the late ninth or early tenth century, and located its origination in Mesopotamia (or at any rate within the lands of the Baghdad Caliphate). By the eleventh century, versions of it were known in Europe and elsewhere, e.g., North Africa. Yassif edited an Ashenazic (Franco-German) version, and a version of Italian background.

39 Italian literature has Giulio Cesare Croce's popularistic tales *Bertoldo e Bertoldino* (in full: *Le sottilissime astuzie di Bertoldo,* with *Bertoldino, figliuolo del già astuto Bertoldo*), of 1606, relating similar exchanges between the boorish, cunning peasant Bertoldo and King Alboino, and featuring in the sequel, Bertoldo's silly son Bertoldino. There also exists a sequel by the abbot Adriano Banchieri [first published in 1620] about Bertoldo's grandchild Cacasenno, who, we are informed by his name, is wit-excreting (which is not the same as being wise). The precedent for *Bertoldo,* in Christian Europe, was the legend *Dialogus Salomonis et Marcolphi.* In *Bertoldino,* Marcolfa is Bertoldino's mother. Antecedents include the late 12th-century German *Salman und Morolf,* the 13th-century Latin *Salomon et Marculfus,* and its German translation *Salomon und Markolf.*

40 I.e., on its upper lip, under the nostrils.

41 Yassif [1984, pp. 239-240].

42 Giving an account of how fiction works, as opposed to reports of real events, has been a subject of inquiry in non-classical logics: Woods and Alward's (in press) is the encyclopedic entry on the subject, catering to logicians, in the new edition of the *Handbook of Philosophical Logic.* There are several theories of the logic of fiction [Lamarque and Olsen 1994, Crittendon 1991, Walton 1990, cf. 1978, Woods 1974]. Pollard [1997] explains: 'Unlike real things, the entities of fiction are incomplete. While the proposition that Napoleon disliked cats is in principle decidable, the proposition that Sherlock Holmes disliked cats is not. On this latter issue, Conan Doyle's texts are silent. Furthermore, according to so-called 'classical' logic, anything whatever follows what is false. If the statements of fiction are taken as false, then what may seem perfectly acceptable inferences made by readers are rendered arbitrary: it would be as reasonable to infer from Conan Doyle's novels that the moon is made of green cheese as it would to infer that Holmes was cleverer than Inspector Lestrade' [*ibid.,* p. 265].

43 Yassif [1984, p. 240, fn. 3.]. Heller [1934] is an encyclopaedic entry on Joshua in Islamic traditions.

44 Prior to the Islamic conquest, Mesopotamia was part of the Sasanian empire, which was Persian, but whose capital was not far away from Baghdad. Non-Iranic people used to speak Aramaic, and this is the language of the *Babylonian Talmud* indeed. By contrast, *Pseudo Sirach* is later, and in Hebrew.

45 This account in the initial part of *Pseudo Sirach* was often found revulsive by rabbinic authorities and others. It relates that a maiden, Jeremiah's daughter (yet in the Bible, Jeremiah is apparently unmarried), was unwittingly inseminated by her father's own semen while bathing in water. Yet, Yassif and others have pointed out that this story of virgin birth may be a reflex of the Zoroastrian tale about the birth of a Zoroastrian redeemer, also born to a maiden with no marital intercourse, and who was inseminated while bathing in the water of a river (wasn't it supposed to flow to the sea?) preserving for one thousand years the semen of Zoroaster. Something similar is related about Buddha. In Christianity, virgin birth has no need at all for the human semen of some prophet. Interestingly, a Syriac and Islamic tradition identified Baruch, Jeremiah's disciple, with Zoroaster.

From *Pseudo Sirach,* it is apparent that its author's rather prosaic mood was making no concessions for making his narratives sublime. Ben Sira's birth is just by mechanical accident. The identity of the author is unknown, but Yassif has shown that a tale about a badly mistreated eagle, related at the end, has no credible function in the economy of this literary work, other than, presumably, in order to convey in allegoric form the author's own misfortunes. On the evidence of a few passages in the book, its author does not appear to have been an endearing character, and he was rather likely to have come through as insolent even in his own lifetime. Yassif shows that his aim was not didactic or ideological, but merely artistic.

46 Ann Mette Heindorff [2006] states: 'Tamgaly is approximately 150 km northwest of Almaty. In Kazakh language the word *Tamgaly* means 'painted, or marked, place' '. There are about 5000 petroglyphs at Tamgaly. Dating is fairly late for such rock carvings: the earlier are dated from the second half of the second millennium BC, and apparently some were being added as late as the beginning of the 20th century. 'For many centuries Tamgaly used to be the place where various rituals and sacrifices took place. Some of the drawings were renewed in the Middle Ages, and used in religious ceremonies until the XI century.' [*ibid.*].

47 The forms of the name of this character include: *Ferydûn, Freydûn, Ferêdûn,* and *Farîdûn* in New Persian, and an earlier form is *Afrîdûn.* In Pahlavic and otherwise Middle Persian, the form is *Frêdôn.* This character already occurs in the *Avesta,* and the Avestic form of his name is *Θraêtaona.* See a presentation of this character from Iranian mythology by Ahmad Tafazzoli [2007]. He explains that the prefix *θri-* 'three' can be detected in the Avestic form of the name, and that 'Θraêtaona may originally have been a patronym from Θrita/Trita, meaning 'son of Θrita', and both Θrita and Θraêtaona can be understood as reflexes of an earlier *Trito (Lincoln [1981], pp. 104-05)'. As to occurrences of the first name *Ferêdûn* and the like: on 5 January 2008, an Afghan-born teenager, Faridon Alizada, was stabbed to death in Erith, in the Bexley borough of London where I live.

48 Tafazzoli [*ibid.*] remarks: 'Ferêdûn's close connection with cattle is noteworthy. His totemic ancestors are mentioned with the suffix *gâw* 'cow' (see ÂBTÎN). This connection may have influenced later reports, for example, his famous mace with a bull's head [*gorz-e gâvsâr*; e.g., *Š*, ed. Khaleghi, I, pp. 71 v. 257, 82, v. 444; Ebn al-Balkhî, pp. 12, 36; Tabarî, I, p. 228; Tha'âlebî, *Gh*, p. 34; Ebn Esfandîâr, p. 58], his nourishment by a cow named Barmâyûn, his being the first to ride a bull on the feast of *dar-e mazînân* [16 Day; Bîrûnî, *Â*, p. 226; Gardîzî, ed. Habîbî, p. 246], his riding a bull while hunting [Ebn Esfandîâr, p. 58], and his fastening the Mâzandars to the hooves of the ox Barmâyûn [*Dênkard*, ed. Madan, 814.11 ff.]'. Yet, for Ferêdûn, association with cattle is not exclusive. Tafazzoli also points out that in Pahlavi and Islamic sources, 'He was the first to tame the elephant and use it in battle and to cross horses and donkeys to produce mules'.

49 Alvappillai Velupillai's text (n.d.) explains Civan's representations in Tamil faith.

50 The image of Europa riding the bull appears on coins of the European Union.

51 The photograph is posted at: <http://www.vroma.org/images/raia_images/altar1.jpg> and is credited to Ann R. Raia [2005]. (The scene from the relief is described at <http://www.vroma.org/images/raia_images/index3.html>).

52 The VRoma (Virtual Roma) Project is a public-funded, US-based project ('VRoma: A Virtual Community for Teaching and Learning Classics',

<www.vroma.org>). The VRoma Project holds the photographer's copyright for this image, and the photographer's permission was kindly granted to me by of the Project Co-Director, Prof. Barbara F. McManus of the College of New Rochelle, in New Rochelle, New York state. Since the photograph was taken in the Capitoline Museums in Rome, I also obtained the museum's permission.

53 Let us consider how he tried to rationalize the myth of the Sphinx. I quote from a review by James Romm [1996] of Stern's translation: '[...] Palaephatus is an odd bird. His peculiar method of rationalizing exotic myths does not fit easily into any of the generic categories established by the Loeb Classical Library [...]. Palaephatus was apparently a contemporary of Aristotle, and thus lived at a time when the body of myths inherited from high antiquity had long been under scrutiny by scientists and skeptics. Plato in the *Phaedrus* shows Socrates playfully attempting to rationalize the myth of Boreas [...]. But Socrates quickly draws back from this sort of approach, anticipating that it would take vast amounts of time and ingenuity to similarly decode all the fabulous elements in the mythic tradition: centaurs, chimeras, gorgons, Pegasuses, and the like. As if taking this *Phaedo* passage as his challenge, Palaephatus set out to [rationalise] all of Greek mythology [...], even when this required performing bizarre, almost ludicrous linguistic or logical contortions. [...] Palaephatus seems to be in dead earnest, since for him the credibility of the whole mythic tradition is at stake. [...] Stern, for the most part, meets this challenge ably. However there are places in which his footnotes could do more to clarify the ways in which linguistic misprisions give rise to Palæphatus' mythic tales. For example, when the victims of a female bandit named Sphinx, who happens to be accompanied by a pet dog, describe the speed with which their attacker moved [...] one can easily see how the image of a winged monster, part dog and part woman, took form. But that image is much harder to derive from Stern's English version, 'She doesn't run, she flies – she and her dog!' It might not have been possible to capture in translation the ambiguity as to whether the Greek *KAI/ ... KAI/* denotes two different creatures or two combined into one, as singular verb suggests; but a footnote would have helped the Greekless reader understand just how natural an error led to the creation of the Sphinx legend. [...]'.

54 Medieval Christian theologians used to refer by *recirculatio* to concepts such as 'lost by the tree – saved by the tree', i.e. the tree of Paradise versus the tree of the Cross. Also Jewish homiletics possessed quite a similar form of association: the retribution principle, מידה כנגד מידה *(middá ke-néged middá)*. It is only a particular case of this principle, that considers retribution in terms of punishment. The latter is the case of Dante Alighieri's use of the notion of *contrappasso.* Cf. the Latin maxim: *Per quae quis peccat, per idem punitur ipse.* In *Exodus* 18:11, Jethro, Moses' father in law, remarks about how apt it was that as Israel was saved, Pharaoh and his army had been defeated 'by that very thing by which they had plotted against them'. Traditional Jewish exegesis maintains: 'By water

they had figured out they would have them lost, yet they were lost by water themselves' [11th-century Rashi's gloss *ad loc.*]. The two episodes referred to are the drowning of the male children in the Nile, and the army being submerged as the waters closed up after the parting of the sea.

55 In Hebrew, *nun* is the name of the letter נ *(N)*, perhaps because it presumably developed from an acronymic pictograph showing a fish, by virtue of the Northwest Semitic name *nun* for 'fish' which we know from Aramaic. In the *Koran,* the prophet Jonah 'is four times mentioned as Yūnus, without his father's name being given, once as Dhu 'l-Nūn [xxi. 87], once [lxviii. 48] as *ṣāḥib al-ḥūt,* 'he of the fish'. This epithet explains also why Yūnus is the only one of the major and minor prophets who is mentioned in the Ḳur'ān; a prophet who is swallowed by a fish naturally attracts attention' [Heller 1953].

56 Of the tale-type 'Born from a Fish' (i.e., type AT 705 of the Aarne and Thompson system), Heda Jason is able to list two Iraqi Jewish oral versions, IFA 11090 and IFA 12020, from the Israel Folktale Archive of Haifa University. See on pp. 45-46 [cf. pp. 31-32] in the Hebrew part in Jason [1988].
AT refers to the classification in Antti Aarne and Stith Thompson's *The Types of the Folktale: A Classification and Bibliography* [editions: 1961, 1964, 1981]. One fundamental tool in the scholarship of the folktale is the classification of Aarne and Thompson indeed. Antti Amatus Aarne (1867-1925) died shortly before Vladimir Propp published his ground-breaking study of the Russian folktales, whose first edition is of 1928, and which paved the way also for story-grammars in cognitive science and automatic story processing within artificial intelligence. (English-language editions of Propp's monograph however only appeared starting in 1958, and the 1958 edition is standard in the U.S.; in Europe on the Continent, one is likelier to use the later, French translation of the Russian second edition). Yet, notwithstanding this chronology, Aarne's work is still essential. Aarne established a system for indexing tale-plots, which was translated from the German and enlarged by Stith Thompson [Aarne and Thompson 1928, 1961], with a recent update by Hans-Jörg Uther [Uther 2004]. Referring to the classification of Aarne and Thompson is the standard fare of the study of folkloric narratives.

57 A medieval folktale about the birth and youth of Joshua is related to the type of the story of Oedipus (a tale similar to which was told, in Greek mythology, about Telephos and Auge), and is unrelated to Joshua's biography as known from the Hebrew Bible. According to the folktale, Joshua's father was righteous and childless. He prayed for a son, but dreamed up that his son would kill him. His wife insisted to learn why he wasn't happy when she became pregnant, and on learning about the dream, she left the baby in the waters of the Nile [cf. Cosquin 1908] – in the same manner that Moses was left there. When he grew up, he became an executioner, was made to execute his father, and the wife and possessions of the executed man was given to him; a miracle prevented incest (not so in oral versions).

It is known that the similar Christian legend about Pope Gregory is closely related. This is discussed on pp. 299-300 in the Hebrew edition of Eli Yassif, *The Hebrew Folktale* [1994]. In n. 29 to ch. 5 [pp. 637-638], Yassif points out that in Gregory's tale incest is not prevented, but parricide is. Yassif cites, for Jewish versions, pp. 68-69 and 199-206 in Edmunds [1985], as well as a paper by Yoav Elstein, contrasting the Gregorius legend in both Christian and Jewish versions [Elstein 1986]. Yassif also cites Cosquin [1908].

58 Germain Bienaimé [1983] is concerned with Moses' association with water.

59 During the Ten Plagues, Moses leaves it to Aaron to bring about the first plague, the one consisting of the transformation of the waters of the Nile, an act of delegation that was homiletically explained as avoiding ingratitude for not having been harmed while a baby inside the Nile's water.

60 Moses' double association with water – his being rescued from the Nile as a baby, and then the parting of the sea he performed – may have contributed to, or even determined, his midrashic (i.e., homiletic-exegetic) association with the blue gemstone *sampirinon* (e.g., this being the stuff of which his rod was made, and this being the material from which he carved the Second Tables). Also consider a midrashic account of Moses' ascent to Heaven from the peak of Mt. Sinai (in order to collect and bring down the Law), which in order to explain how he could move in the air, describes him as swimming in the air like a *kolyas* (κολίας) fish does in the water. (For a tale about Moses seeking to be transformed into a fish, see below). The likening to a *kolyas* fish in the context of Moses' ascent to Heaven from the top of Mt. Sinai is likely to reflect a creative *ad hoc* explanation on the part of a specific late-antique homilete, rather than the transmission of, say, a totemic concept. Yet, who could rule out the possibility that the very image (however applied) was in existence in antique culture(s) in the given geographic environment? The traditional date of both Moses' birth and death is the 7th of Adar, a month of the Jewish lunar calendar whose Zodiac constellation is Pisces.
Why the *kolyas,* of all fishes? Zoologist Menachem Dor died at 100; his last book was Dor [1997], on the fauna in the Bible and in early rabbinic sources. In Dor's entry, on p. 176, for the the fish called in the early rabbinic sources the 'Spanish *kolyas*', which he identifies as the Spanish mackerel, Dor stated: 'It is a close relative of ther tunny, and like the latter, it is a good swimmer. In *Tanḥuma, Ki Tissa,* 32, it is written that Moses swam in the sky like a *kolyas* in the sea'.
Being transformed into a fish is proposed by Moses (along with other candidate mutations) in *Midrash Tanḥuma, Ki Tissá,* 32. In Rella Kushelevsky's [1995] *Moses and the Angel of Death,* on p. 174 such midrashic passages are discussed, in which Moses implores the Lord to let him enter the Promised Land, even if it takes being transformed into a bird (according to *Deuteronomy Rabbah* and to the *Jellinek-A* version), or into a fish *(Jellinek-A).* Kushelevsky analyses this as instances of Motifeme 'Hero offers reductive transformation of self'. In *Jellinek-A*

[Kushelevsky, p. 174], this is the request of being transformed into a bird: 'Lord of the worlds, implored Moses, let me at least, by the power of the Ineffable Name, fly like a bird in the air', and here is his request to be transformed into a fish: 'Or make me like a fish, with my arms spread like two fins and my hair turned into scales, so that I may leap over the Jordan and see the Land of Israel'. By contrast, in the lore about locomotion in fluid like a *kolyas* fish, both swimming like a fish, and flying like a bird are involved.

61 David Ḳimḥi (b. 1160?, d. 1235?), in his gloss to the given verse from *Joshua,* quoted that homiletical passage, and added: 'That is to say that the mountain quaked on the day he was buried'. Rashi (b. 1040, d. 1105) stated in his own gloss *ad loc.:* 'It informs that the mountain quaked on them to kill them, as they had not eulogized him befittingly'.

62 That bull-riding in the given story has any lunar connotation cannot be proven, unless such an association in Iranic mythology is found. (That in Aegean culture the bull had lunar connotations is not relevant.) We do find instead an association of bulls (and horned, hoofed animals in general) with the Sun (thus, an opposite of the Moon), in western Inner Asia, which in turn has connections in ancient Iranic culture. This importance of the Iranic link to our story is because of the relevance of Ferêdûn, which in turn is because of *Pseudo Sirach*'s likely dependence on Islamic lore when it comes to Solomon and Sheba (so Islamic lore is relevant), and because Ferêdûn became associated in Islamic lore with more than one biblical character (albeit we don't know that Joshua was among these).

63 Gerd Fritz [2006], reviewing Arie Verhagen's [2005] *Constructions of Intersubjectivity,* stated on p. 592 that in Verhagen's Ch. 2, his 'strategy is to start with items of vocabulary that are good candidates for an analysis in terms of intersubjectivity and to generalize from there. Expressions like *little chance* or *barely half full* tend to be used differently from truth-functionally closely related expressions like *a small chance* or *almost a half full.* By a detailed analysis he shows that they are routinely used to signal a negative or sceptical attitude of the speaker and thereby to invalidate a relevant positive proposition the hearer may entertain. [...]'

64 Women were likelier to have heard a tale, rather than to have accessed manuscripts directly.

65 An instance of cultural appropriation of visual representations made by the Other on the part of viewers, is as follows. The Portuguese, during their conquests on the west coast of India, destroyed local temples and the sculptures found there, which portrayed deities. Yet, those same Portuguese spared sculptures with the three bodies of the Indian Trimurti, because they identified them as representations of their own Trinity. The resemblance is not casual. The three-headed visual representations in Europe and in India are correlated, and Pettazzoni [1946] discussed the historical background of the European version.

66 Rachab looms large in the biblical narrative of Jericho. See *Joshua,* 2:1-21 and 6:17-25. Rachab is not mentioned in *Pseudo Sirach,* from which the tale about Joshua riding a bull is taken.

67 Elsewhere, in work in preparation, I analyze what arguably are psychologically motivated typos. When I typed, in the present article, text about the Roman relief from the Capitoline Museums, at first I typed: 'One of the scenes in those reliefs shows a hero riding a bull beside goddess Italia, with a city packground'. The letters *p* and *b* are not near each other, on the keyboard I was using. Arguably I mistyped *packground* instead of *background,* because there were all those visual things packed together in the relief. This is an example of how in a domain – typos – which is typically ascribed to chance, sometimes a causal link can be detected. The odds still apply: you would usually *not* mistype.

68 An example of chronological conflation in pictorial representations from the Middle Ages is provided by an illustration in the Holkham Bible Picture Book (Holkham MS 666 from the British Library), from the 14th century, a manuscript in which the illustrations are captioned with notes in Anglo-Norman French. We are concerned with a particular image, reproduced in black and white in Fig. 9, and showing Noah in his Ark with extended arms, and looking to his right (i.e., leftwards, in the picture). In his left hand, he is holding a crow he is about to release. A dove stands on his right hand. Even more to the left in the picture, facing that dove there is an image of the same dove, flying back, and carrying a little branch in its beak. Under the Ark, under a wavy line with a blue contour, hoofed animal legs point upwards, the pale brownish animal having drowned. More to the right, a pale pink drowned person is floating. Further down in the picture, another human corpse is seen, as well as (further to the right) a drowned dapple brownish horse, its eye perhaps about to be gouged: the crow is showed beaking the head of that horse. This is the same crow that appears above, also in the rightmost part of the picture, but at the top, while about to be released by Noah. In contrast, on the left side of the picture, at the bottom, below the dove at the top, about to reach the Ark while holding an olive branch in its beak, that same dove is shown while she detaches that branch from a tree on an emerged spot of land. Pale bluish water alternates with darker blue wavy lines, but the emerged spot of land is painted dark brown, with a pale brownish green, tiny, slender tree on top of the tiny island.
Within that same picture, a sequence of six moments in time is conflated: (1) the outcome (floating bodies) of the drowning of living creatures not in the Ark, (2) the release, by Noah wearing a hat with a red front, of the crow, whose mission's eventual outcome was disappointing, (3) the crow behaving like a crow, with the extra-biblical detail of this crow feeding itself on a carcass, (4) the release of the dove, (5) the dove finding emerged land with an olive tree on it, from which the dove detaches a branch, and (6) the return of the dove to the Ark with a branch of olive in its beak. A medieval or later onlooker could be expected to possess the

conventions for deciphering the picture, and decomposing it into the different moments of the given narrative depicted.
Consider the crow perched on the head of the drowned dapple horse. Perhaps the horse's eye is already inside the crow's beak, with the ocular nerve dangling out of it. There is a contrast (reflected in the symmetrical positioning inside the picture) between the dove being portrayed as dutiful, and the crow being portrayed as seeking revulsive advantage from the universal calamity. The original biblical text does not imply that much.
In terms of **chance,** the dove achieving for Noah what the crow did not, is clearly conditioned by the crow's mission having preceded the dove's. The waters receding were a concomitant process with the sequence of the crow, then the dove being sent out of the Ark. Therefore, the positive outcome is obtained when it is the dove that has been sent out on perlustration.

69 *Time-slicing* is a concept that occurs in a number of different disciplines. It can be found in computer operating system design; in cinematography; in synchrotron radiation from physics; in power engineering; and so forth. In a computer's operating system which, on a serial machine, has to simulate the parallel execution of a large number of programs, this is achieved by means of time slicing: at each time slice, a small number of instructions are executed for each program, one at a time. If the strategy is that a fixed number of instructions is to be executed for each program at each time slice, longer programs will take more time slices. Other strategies are possible.
In telecommunications, time slices (i.e., burst, or time slots when the data are transmitted) enable a reduction in power consumption, e.g., of mobile receivers, because when the relevant data are not available, the front end (i.e., the receiver) is switched off, and later on it is informed when to wake up, when the next burst (intended for it) is expected.
As to special effects in cinematography and computer animation, the Wikipedia entry for 'Bullet time' (<http://en.wikipedia.org/wiki/Bullet_time>, accessed in March 2007), remarks that 'Technical and historical variations of this effect have been referred to as time slicing, view morphing, flo mo, mort temps and virtual cinematography'. It is 'a computer-enhanced variation of slow-motion special effects used in some recent films and computer games. It is characterized both by its extreme permutation of time (slow enough to show normally imperceptible and un-filmable events, such as flying bullets) and by the ability of the camera angle – the audience's point-of-view – to move around the scene at a normal speed while events are slowed. This is almost impossible with conventional slow-motion, as the physical camera would have to move impossibly fast; the concept implies that only a 'virtual camera', often illustrated within the confines of a computer-generated environment such as a game or virtual reality, would be capable of 'filming' bullet-time types of moments'.

In beam physics, a report in the *Advanced Light Source (ALS) News* [ALS News 2000] stated that '[i]n early 1996, Alexander Zholents and Max Zolotorev of Berkeley Lab's Center for Beam Physics proposed the laser time-slicing technique as a way to achieve effective bunch lengths in the femtosecond range. At the heart of the proposal was the use of a high-power, femtosecond laser synchronized with the electron bunches so that a pulse of laser light passed collinearly with an electron bunch through an undulator or wiggler. The high electric field of the shorter laser pulse modulated a portion of the longer electron bunch, with some electrons gaining energy and some losing energy. Subsequently, when the energy-modulated electron bunch reached a bend magnet (or other section of the storage ring with a nonzero dispersion), a transverse separation occurred. A collimator or aperture selected the synchrotron radiation from the displaced bunch slices'. Later on, according to the results of Schoenlein et al. [2000], a Berkeley team 'generated 300-femtosecond pulses of bend-magnet synchrotron radiation at the Advanced Light Source (ALS) with the aid of a laser 'time-slicing' technique. Their proof-of-principle experiment demonstrates that this technique is a viable one for producing ultrashort pulses of x rays. An ALS bend-magnet beamline will soon be commissioned that will be dedicated to time-resolved x-ray diffraction, EXAFS, and other techniques capable of probing the long-range and local structure of matter on a femtosecond time scale' [ALS News 2000].

Time-slicing is also applied to archaeometry: ground penetrating radar (an earth survey method from geophysics) is put to good use for intensively recording an area, slicing the data into images, possibly animating the images by time-slice (e.g., to show ancient cultivations or archaeological features).

Within artificial intelligence, *slice* comes from the terminology of *naive physics.* It was introduced in Patrick Hayes' [1985] *ontology for liquids*. The latter is a naive physics model that can handle phenomena in fluid dynamics as well as an important class of basic phenomena in common-sense perception of everyday life. This involves spatio-temporal *histories* of objects. Containment, filling and emptying [*ibid.,* p. 97] are dealt with in the context of Hayes' Ch. 7, titled 'Change and Movement: Histories' [*ibid.,* p. 90 ff]: 'A history differs from a situation in being restricted spatially and extended temporally: it is a connected piece of space-time in which 'something happens', more or less separate from other such pieces. Histories, unlike situations, have a *shape:* much of [Hayes' treatment is] devoted to ways of describing their shape. Examples of histories include the inside of a room during the afternoon, a horserace and the pouring of water from one cup into another. The idea is that a history shall contain an event, isolating it temporally and spatially from other events. We include the special case in which nothing happens at all. A *state* [...] is an instantaneous 'slice' of a history at a certain time-instant. [...] If *h* is a history and *t* a time-instant (we assume a global timescale of some sort with an inequality defined on it), the *h@t* (read *h* at *t*) is the 'slice' of *h* at *t*. This is a state, that is, a spatial entity at a particular time' [*ibid.*, p. 90].

70 The subnarrative about the converted Rachab's marriage to Joshua is related in the *Babylonian Talmud,* tractate *Megillah,* 14b, as well as in *Pesiḳta de-Rav Kahana,* 13, 110. It is related that eight prophets or priests were descended from that union, including Jeremiah and Ezekiel.

71 It is not accepted by all late antique and medieval exegetes that Rachab was a prostitute. As her described actions are as an innkeeper, an often expressed interpretation is that she was a caterer, one providing food, by deriving *zoná* from the same lexical root as the noun *mazón* 'food' and the transitive verb *zan* 'to feed', rather than from the same lexical root as the intransitive verb *zaná* 'to fornicate' (etymologically and metaphorically also: 'to display a wayward/loose behaviour'). The more innocent explanation is the explanation that I and my classmates got at primary school, while studying the *Book of Joshua.* In the 11th century, Rashi (Troyes, Champagne, 1040-1105: his biblical and talmudic commentaries are the Hebrew *glossa ordinaria*) referred to a late antique Aramaic translation, and stated: ' "A woman [being a] *zoná*" – Jonathan translated it: *pundeḳéita,* she who sells foodstuff'. This interpretation was prevalent, because of the spread of Rashi's commentary exceeding the availability of the other commentaries. In Provence, the exegete and grammarian David Ḳimḥi (or Radaḳ, b. 1160?, d. 1235?) provided a more elaborate explanation: ' "A woman [being a] *zoná*" – as per its [literal] sense [i.e., a prostitute]; or [otherwise] she who sells food, as per Jonathan's rendition: *pundeḳéita,* [...] Yet in truth, the intention of Jonathan [was that she was] really a prostitute, because he is used to translate [almost] everywhere any [instance of] *zoná* as *pundeḳéita,* [...] even though he translated a few instances as *nafḳat bara* ['she who goes outside'] [...] and his intention [is that] a prostitute is like an innkeeper, available for all'. Women who were innkeepers were held to be disreputable, at least in late antique times.

A later medieval biblical exegete, Joseph Caspi, from 14th-century Provence, interpreted: ' "*zoná*" – as per its plain sense [i.e., a prostitute]; besides, most [often] innkeeping is associated, because they used to do that as a cover for recondite [disreputable] activities'. Joseph Caspi, known in his milieu as Don Bonafous de Largentera, was a philosopher and a grammarian, as well as a biblical commentator. He was born around 1280 – previously, the date was taken to be 1297 – at Largentière, and died at Tarascon in 1340. Hebrew *kaspi,* 'made of silver', here means 'one from Largentière'. Barry Mesch [1975] dealt with Caspi. Basil Herring's [1982] is a critical edition of his philosophical Bible commentary. Cyril Aslanov [2001] showed, based on *Sharshoth ha-kesef,* a Hebrew-Provençal dictionary by Caspi, the latter's importance also for documenting Provençal.

72 Moreover, the subnarrative about Joshua marrying Rachab also claims that some prophets were descended from them. During the early 1960s, there was a humor-

istic broadcast on the Israeli radio. It was set in Joshua's own days, but it was structured as a debate among modern panelists, the subject being: should the planned wedding between Joshua and Rachab take place? Ephraim Kishon has published a humoristic piece about Moses's contemporaries expressing views on Jethro, his influential father-in-law. (Jethro is the one who on seeing the people queuing, persuaded Moses not to micro-manage, and to establish a judiciary cum administration to whom to delegate, lest he would succumb to exhaustion. The exhortation to the leader to delegate what he could not handle well in person, is also found in Plutarch's treatise on *The Precepts of Statecraft,* 15.811B-812F.)

73 The textual evidence is that Rachab tells the two spies that she and the other people are afraid. But how afraid is she? The Roman relief, if believed to show Rachab, would be showing the possible *effect* of mood or emotion, not any direct display of these. (Actually, as we believe that the relief was a pagan scene with a goddess, she is unlikely to have been represented as being afraid). Displays of emotion are a cue for narrative interpretation, in Dyer's [1983] BORIS narrative-understanding program. Causality is assumed. Aetiological tales also arise sometimes from displays of emotion; e.g., in Islamic traditions about Jonah:

> Yūnus serves Allāh in Niniveh. Why was he enraged? 1. He lived at a time when the 9½ tribes were already made captive. It was him king Ḥizḳia sent to the exiled, though there were five prophets in Israel at the time. Yūnus is angry because the mission is entrusted precisely to him. 2. Yūnus is angry because his mission is so urgent that he has not even time to put on a shoe or to mount a riding animal. 3. Yūnus is enraged because he sees that his threat to Niniveh has not been fulfilled so that he appears to be a liar [Heller 1953].

74 This was also the case of North African costume, as pointed out by Labelle Prussin [2006], whose paper traces the impact of Jewish artisans on West Africa:

> From the hands of these Tuareg [...] blacksmiths came the uniquely configured, spectacular triangular pendant [...] that was once the pride of every Tuareg woman [...]. Its design is no more than an elaborately magnified version of the triangular motifs of the Moroccan fibulae, albeit in a simplified geometry and technology. With the shift in costume from the traditional North African draped toga to a tunic, fibulae were no longer functionally requisite, so they were merely inverted and turned into a singular pendant. The fastening clasp was abandoned in favor of the simple loop, [...]

75 Philipp Fehl [1974] discusses where the equestrian statue of Marcus Aurelius was located in the Middle Ages. Dieckmann and Seznec [1952] are concerned with a controversy between Diderot and Falconet about the horse of Marcus Aurelius.

76 In 1538, the original was placed in Capitol Square (Piazza del Campidoglio) by Pope Paul III.

77 This coin is made of so-called Nordic Gold, and its diameter is 24.25 mm.

78 Between the 12th and the 16th century, there used to be a genre, called *Mirabilia urbis Romae,* of guides for pilgrims, which took interest in artistry [Schlosser Magnino, edn. 1996, pp. 54-55, 600]. Variants evolved. One version was translated by Francis Morgan Nichols [1889]. In the early modern era, modern guides developed. The classic of the guide genre was the inventory of art on public display in Rome, the *Studio* first published in 1674 by Filippo Titi (b. ca. 1640, d. 1720 or 1721), and then reprinted in several revised editions that were well received by the public [Titi 1763 = 1978]. In the second half of the 18th century, it was superseded by Giuseppe Vasi's *Itinerario istruttivo* [Vasi 1763], often reprinted until 1820, and eventually thoroughly rewritten by Antonio Nibby, whose own version is of 1821 and had French and English editions, then by Valentini, and finally by Porena in 1886, was in common use among tourists, pilgrims, and the cultivated public until the end of the 19th century [see Prinzi's preface to Titi 1978, and Schlosser Magnino, edn. 1996, pp. 544, 602].

79 My translation (emphasis added), p. 55 in the 1996 Italian edition of Julius Schlosser Magnino's [1996] *La letteratura artistica.*

80 One aetiologizes, sometimes, also when telling a personal narrative (e.g., 'I guess I did it because...', or less tentatively stated reconstructions). Autobiographical accounts are categorized in folklore studies as a 'memorate', whereas had the narrator ascribed the narrative to another person (even should it have been an autobiographical account of the latter), then the narrative would count as a 'fabulate'. Linda Dégh and Andrew Vázsonyi [1974] discussed these concepts. Lauri Honko [1964] is concerned with memorates in connection with folk beliefs.

81 Vern Walker [2001] reviewed jointly the books by Shafer [1996] and Kadane and Schum [1996].

82 See in note 4 above.

83 Gadda's novel [1957] was first published in instalments, in 1947, in the literary magazine *Letteratura.*

84 At the site <http://www.arts.ed.ac.uk/italian/gadda/Pages/resources/babelgadda/babeng/thatmesspocket.html> extracts are posted, from Chapters 1, 2, 4, 5, 6, 9 and 10 of Weaver's English translation of Gadda's novel. The website is that of *EJGS BabelGadda,* of the *Edinburgh Journal of Gadda Studies.*

85 At the site <http://www.arts.ed.ac.uk/italian/gadda/Pages/resources/babelgadda/babeng/deluccamess.html> the partial translation by Robert de Lucca is posted, in *EJGS BabelGadda.* One can hardly imagine a more difficult text to translate, than this novel by Gadda, because of its idiosyncratic language.

Ephraim Nissan, Goldsmiths College, University of London

Giuseppe Padovani

Chance and Sense of Action

Mais cela est artificiel. Car, partout et toujours, l'accident jouera ou a joué un rôle que Cournot s'obstine vainement à juger transitoire, et, somme toute, insignifiant, et destiné à ne pas laisser de trace profonde.

Gabriel Tarde

Abstract:

The study of chance sets some problems to the epistemology of sociological action theories bound to the role carried on by the observer. As a matter of fact the purpose of an action would remain ambiguous to an observer when depending on the irruption of chance, by these cases Aristotle adopts a position that identifies the point of view of the actor to the one of the observer. Theories about olistic models of action embrace this position: the acceptation of a system intentionality would give access to an epistemology recognizing chance, its unforeseeable irruption, by an objectivistic perspective.

Keywords:
chance, epistemology, intentionality, observation, social action.

In the literature dealing with chance we find discussed and quoted more or less the same examples, even if many are the chances may happen to a man and to men as well. So dealing with chance, or fortune, in The Metaphysics Aristotle quotes the example of somebody digging to plant a tree who finds a treasure [cfr. Δ 1025 a, 15]. This example about what may happen by chance to a man, states anyway many problems to social action theories.

How comes possible to assert that somebody find by chance a treasure rather than saying that he dig up a treasure? How comes possible to impute this intention rather than that one? We may believe with a certain grade of reliability that commonly digging doesn't mean to search or to dig up a treasure: the word 'treasure' means something rare! Then is rather than unlikely to find a treasure by digging a ditch, as possibly it may happen.

A classical example is undoubtly the one of the falling of a tile on the head of somebody passing by: example reprised by E.E. Evans-Pritchard and commented by Lévi-Strauss. For Azandes bad luck depends on witchcraft acting in sintony with natural forces; so the fall of a wooden beam floor together to the witchcraft is cause of the death of the man who was under it.

Sorcery, or magical thought, or witchcraft are not phisical causes of the fall of the floor, that would have fall down anyway, but are responsible of the fact that the floor fell in a specific moment when underneath there was somebody. In other words, witchcraft explains *why* events are bad to man and not *how* they happen [E.E. Evans-Pritchard, 1936].

It may be said that in the sphere of social life what comes to be a problem for individuals is not the happening by chance but its rationalization, explanation. By the fact, chance challenges individuals intelligence, the order of things and of their course, even for its esthetic taste; order anyway has to do with proportion, harmony, consonance of things, theirs being in order-to.

> La vraie question – writes about the french anthropologist – n'est pas de savoir si le contact d'un bec de pic guérit les maux de dents, mais s'il est possible, d'une certain point de vue, de faire 'aller ensemble' le bec de pic et la dent de l'homme (congruence dont la formule thérapeutique ne constitue qu'une application hypothétique, parmi d'autres) et, par le moyen de ces groupements de choses et d'être, d'introduire un début d'ordre dans l'univers; le classement, quel qu'il soit, possédant une vertu propre par rapport à l'absence de classement. [C. Lévi-Strauss, 1962, p. 16]

In any way, in absence of any rational explanation or supposed being so, is so far better research how to come to an agreement for the tile and who is passing by underneath, for digging and the treasure. The research of similar agreements may seem unreal, and still any order, even a magic one, tends to restore a place for every thing.

> Mais n'est-ce pas que la pensée magique, cette 'gigantesque variation sur le thème du principe de causalité', disaient Huber et Mauss, se distingue moins de la science par l'ignorance ou le dédain du déterminisme, que par une exigence de déterminisme plus impérieuse et plus intransigeante, et que la science peut, tout au plus, juger déraisonnable et précipitée? [C. Lévi-Strauss, 1962, p. 18]

The Azande magic and witchcraft example of Evans-Pritchard consent to make agree the tile and the head. Magic opportunely questioned makes possible to get back to an order by what happens, it is a type of knowledge nearby the scientific one. Magic is not much the manifestation

> une appréhension inconsciente de la *vérité du déterminisme* en tant que mode d'existence des phénomènes scientifiques. [...] Il y a plus. Non seulement, par leur nature, ces anticipations peuvent être parfois couronnées de succès, mais elles peuvent aussi anticiper doublement; sur la science elle-même, et sur des méthodes ou des résultats que la science n'assimilera que dans un stade avancé de son développement, s'il est vrai que l'homme s'est d'abord attaqué au plus difficile: la systématisation au niveau des données sensibles, auxquelles la science à longtemps tourné le dos et quelle commence seulement à réintégrer dans sa prospective. [C. Lévi-Strauss, 1962, p 19]

The example of the tile is drawn on many times in its works by A.A. Cournot to stress that 'nulle connexion, nulle solidarité, nulle dépendance entre les causes qui amènent la chute de la tuile, et celles qui m'ont fait sortir de chez moi pour porter une lettre à la poste'.

A a matter of fact what type of relation might be between the order of causes that makes fall a tile and the order of causes involving my behaviour? A tile may fall from the roof for many physical causes easily recognizable; likewise someone's behaviour may be justified with regard to reasons pushing him to carry out that certain action.

What stays still unexplained, what must be explained, is the fact that the tile falls right at the given moment when he was passing by that spot, it is the happening of this certain event. The problem comes out from one to the other's *solidarity* between these two sequences of independent causes, even when the tail is not hitting him.

The sequence of events setting up the fall of the tile or the wooden beam floor has nothing to do with the one related to someone's behaviour, it doesn't determine it at all, and still it happens that the individual is hit by it.

If happens to somebody to watch someone else, X, busy at digging a ditch in the ground, let's say at the back of his house, and sees that after a while lifts up from the ground a case full of gold coins, may very probably believe that X have digged up a treasure formerly buried in that spot by others or himself.

He may therefore describe the action he watched attributing to X the aim to seize the treasure there buried and believe that he will put the plant at the place of the treasure.

Now, if the action of digging stayed devoid of the indispensable complementary components to be described as an action, could not be described as an action. Why someone's digs? Who's digging? is the answer to these questions allowing to recognize an intentional action. Nevertheless on E. Anscombe's opinion not in all cases the answer allows us to recognize an action as intentional: a same action could indeed be intentional under a description (the peasant digs a ditch to plant a tree) i.e. 'the description under which he doing something', and not intentional under an other (he finds a treasure). The action may be described in two ways: as consequent from an intention of the agent and as consisting of an unintentional process. It is intentionally that the peasant digs to plant a tree? Yes, he performs all the necessary gestures to dig a ditch in the ground. Is it intentionally that the peasant finds a treasure? No, it is a coincidence, lucky in this case.

An observer, as besides anyone of us, is witness of situations belonging to a familiar world.

He 'sees' what X is doing and 'knows' what X does, i.e. digging. Anyway for him the *meaning* of the action is not so much cut down to digging, asking himself why X is doing what is doing? The observer believes that who acts may act by an aim, by a reason, a goal, as planting a tree, as says Aristotle himself. Anyway when he sees him lifting up a treasure, how can he state X *finds* a treasure?

We believe this is a common problem not only to an outer observer, a researcher on the field, but also to theories aimed to the reconstruction of reasons of action. In general terms, in fact, these last identify the viewpoints of the agent and the observer, as Aristotle, or, in compliance with instructions contained in the works by Wittgenstein, renounce to consider the intention only as an interior state of the subject agent.

According to M. Weber either for the observer or the social actor, the action reveals a '*subjective sense'* that must be understood and interpreted, because it is peculiar to the acting subject.

Even for social scientists action carries a meaning, because who acts has 'good' reasons to act in the way he acts and, accordingly, for social scientists it is essential searching for, understanding and interpreting.

How to get to know those reasons? It is enough to assume as starting point that 'Ce qui distingue l'action de tous les autres événements, c'est précisément l'intention'. Ricoeur in this passage shields intentionality of action, as far as it is an event.

So that towards the analytical philosophy of action upholds that distinction between action and event is not a matter of verbs grammar:

> Les actions sont certes des événements, pour autant que leurs description désigne quelque chose qui arrive, comme le suggère la grammaire des verbes, mais nulle grammaire ne permet de trancher entre des verbes qui ne désignent pas des actions, tels que 'trébucher', et des verbes qui désignent des actions, tels que 'frapper', 'tuer'. En ce sens, la distinction entre faire-arriver et arriver.... tombe à l'intérieur de la circonscription des événements. C'est l'intention qui constitue le critère distinctif de l'action parmi tous les autres événements. [P. Ricoeur, 1990, p. 94]

The problem raised hits the question put by us without working it out. The intention is either a quality, 'adverb', of the action, i.e. the way certain actions are carried out, or is an act of intending ('acts of intending', Anscombe) of the subject to whom the action be ascribable. Is the intention, the act, upholds Ricoeur, allows to tell an action from a simple or sheer event, that is impersonal.

What is that allows an outside observer to tell a whole of gestures constituting the action of digging sometimes as expression of intention of planting a tree, sometimes as digging up a treasure and sometimes of finding a treasure?

Let's stop on the verb 'to find': '...and he finds a treasure'

What makes possible to Aristotle to say that someone digging 'finds a treasure?' The question hides a problem of epistemological order. Aristotle expresses a proposition with the verb at the third person, when expressing at the first person, he could have written 'X' says: I have found a treasure as 'I find a treasure'. In these cases the problem would not put himself in the same way, as a matter of fact he would admit he found a treasure just because he himself found a treasure.

The use of the same verb at the first person is different from the one at the third person.

Still, having made use of the verb at the third person since is not him who found the treasure, it takes to mean he may make use of it as 'he knows' that the digging one found the treasure, which is something well known.

He could even have acted following the folk belief that treasures are to be found, mostly by chance, digging, just think about Aesop's fable 'The peasant and the Fortune'.

No matter how it is, it might be possible to uphold somebody has found a treasure if the one who finds it doesn't tell he found it? It is possible to infer it from observing the scene of the finding? It is possible to observe such a scene as the finding? By our case what is 'to know' that a treasure get found if not knowing that happens, that the finding takes place?

'I find a treasure'

By this case, if Aristotle had put such an expression, we wouldn't put it in the same way, because the meaning of the predication changes, the use of verb 'to find' changes at the first person. When I say 'I find a treasure', I myself stand surety for the finding, I 'know' it.

I 'know' it because I was not seeking a treasure; otherwise I would have found what i was seeking, i.e. the treasure. In fact I was not seeking a treasure, I was digging to plant a tree. Again, as outer observer, how can I say 'he finds a treasure' when I don't 'know' he was seeking a treasure? How comes it possible for me to say 'he sought'? If the one interested in it or somebody else would not have told me or report that directly to me?

The verb 'to seek' is an intentional verb, while the verb 'to find' it is not anyway. The verb to find combined to verb to seek is without any doubt intentional, while by oneself it is not.

The clause including the verb 'to find' implies a relation with the verb 'to seek' linked to it by an inner logic: One finds what he was seeking. Once found what we were seeking, nothing get found by chance; while as we find something we were not seeking at all, well in this case it happens by chance, especially talking about treasures.

The use of verb 'to find' is different if linked and coordinate to verb 'to seek'. When in relation to the verb to seek, carries explicitly or implicitly an intention 'finding comes from seeking' and, therefore, the finding is not left to what happens by chance; if on the contrary it is not explicitly in relation to 'to seek' it doesn't carry by then the lack of an intention in acting but the irruption of a happening: it happens right to find. And you do find a treasure by chance!

The intention acts as connector and coordinator of the use of 'to seek' and 'to find': a same intention relates between themselves to seek and to find; likewise what is to seek and to find have to be thing or the same person. The subjects might be different, but sharing same intentions, but it could not be different the thing or person to be sought and, therefore to be found.

When we go into the question here above, it is possible to believe that we have something to do with two ways of knowing: one relevant to the intention and the second relevant to the remark of what takes place, of what happens in the action. Anyway, as we will mention in what follows, they are to be re-taken by a higher level.

The grandchild may help her grandmother to seek out the glasses she lost!

Anyway may even happen that the grandmother say to her helping grandchild: 'Don't worry, as far as I seek my glasses, I never find them. It is only when I don't seek, that I find them. They will come out by themselves!'

The grandmother knows by experience that as far as 'to seek' and 'to find' used together be intentional verbs, sometimes they act as they were not and, therefore, can be used as they were impersonals.

You don't seek things only when are lost, but they make themselves found when are not sought, it is just like the glasses, when not sought, make themselves found.

The two verbs may be used combined and disjoint in situations and happenings different between them.

Let's compare these two propositions:

it is possible to find without having sought, if one knows he is seeking;

it is impossible to find what someone doesn't seek and that doesn't know he is seeking.

Who digs may find a treasure he hadn't sought at all, so the grandmother may find the glasses even if she is not seeking them, but neither in one nor in the other case what get found is re-cognised as a hidden or lost object. Therefore the verb to find, even if in a roundabout way, is an intentional verb. If on a hand it is possible to uphold by V. Descombes that

> Un verbe intentionnel n'est pas intentionnel à l'état isolé, il l'est par son intégration à un système des verbes intentionnels. Parmi ces verbes intentionnels, on distinguera des verbes strictement intentionnels, comme 'chercher', et des verbes occasionnellement intentionnels, comme 'trouver'. On pourrait dire que les premiers ont une intentionnalité *à plein temps*... Les seconds possèdent une intentionnalité *à mi temps*, et c'est ce qui leur permet d'assurer à l'ensemble des verbes intentionnels le contact ou la connexion positive avec la réalité, un contact qu'il faut concevoir comme une relation mutuellement réelle entre deux êtres bien individués. [V. Descombes, 1996, p. 88]

On another side, these verbs create trouble when are used joined together the expression 'by chance' (I refer, partly, to the second clause of the quotation from Descombes). As a matter of fact, the use of the verb to find requires that what is found be recognized, i.e. known, even when is not sought. Indeed the intentional activity is not yet a transitive act, 'une relation réelle', the grandmother may seek the glasses and finding them not, but would it be possible to find without seeking them? I can tell I find a treasure as 'I know' what a treasure is like; how the adverbial specification 'by chance' modifies the meaning of the clause? It doesn't act only contrary to a intention (to seek), but even to an impossibility (it makes impossible to tell that something be found as it is unknown, when by chance means what totally new and therefore unpredictable), when it doesn't act contrary to, the chance, is regarding what possible.

Got to this point the following remark is established: the backgrounds of the actions of finding, so to say, by those grandmother's glasses got lost and the treasure has been buried, however different by breadth and by characteristics, are given and finished, they set up a given and knowable reference frame. In fact the grandmother may tell her grandchild, with more certainty than someone digging may have, that her glasses will make themselves found, that seeking them is useless! The grandmother seemed to take up an unlike viewpoint from the one of Aristotle, by which the chance is arranged in an other way, closer to the one belonging to the probability game.

For Aristotle things are or may become or happen out of necessity and by accident (moreover, otherwise by chance). Things that are and happen out of necessity or by accident are referable to causes, the relation cause-effect expresses a logical connection of the kind 'if X, then (or often, sometimes) Y'; the possibility, the act/power ratio, express the ontological link of them.

What is by accident may even have not been, to what that is by accident belongs and/or belonged and/or will belong to the possibility to be and not to be. Since what is by chance is a contingent being to this is worth to the possibility to be and/or not to be.

Furthermore it distinguishes chance from luck. What is by chance has undoubtedly an ontological statute even when its logical and epistemic connection is unknown. In the epistemic scheme chance is not a residual concept; it is essential to the economy of the scholarship: moreover when removed as the Megarians believed, we would jump to the absurdity of the deterministic side.

By chapter 9 of *ΠΕΡΙ ΕΡΜΗΝΕΙΑΣ*, Aristotle remarks that believing that everything happen out of necessity would be 'absurd' as it would be like denying the evidence of the becoming and of the change, i.e. the possibility of the contingent.

In the Metaphysics [VI, 2, 1027, a 15-17] reaffirms 'beyond what is forever and mostly, there is what happens by chance and by accident'.

In the *Physics* Aristotle uses two terms to specify chance *τύχη* and *αὐτόματον*, that

> explicitly means – when the greek language is spoken by the episteme – the irruption of the becoming (*μάομαι*) by something the one protagonist and responsible of the irruption; therefore it irrupts 'by himself' ... (*αὐτό*) – and is not and has nothing before irrupting. [E. Severino, 1979, p. 24].

Chance is something irrupting by itself, its cause can be nothing but indetermined and, therefore, unknowable. All it is and becomes has to do with principles and causes in the case of a voluntary action, the principle of action is in the agent, 'depends by him acting, or not acting, the actions whose principle is in it' [Eth. Nic. 1110a, 16]. The principle of voluntary action is in whom that acts insofar as he decides with respect to what is in his power to realize ('deciding is given in cases mostly happening, those whose end is still unknown, and those by what...there is somewhat undetermined', [Eth. Nic. 1112b, 7] this is the reason why luck is connected to acting). In fact it is not possible to decide about everything 'We decide on the contrary on things depending on us and which are realizable: These are the remains' [Eth. Nic. 1112a, 30].

Man can not decide on what happens by nature, out of necessity or by chance, althought luck be proper of acting; moreover what is by chance is something irrupting by itself. Man can not decide on 'things springing from chance, as for instance on the discovery of a treasure', even though this happen.

The cause of the discovery of a treasure can not be ascribed to man's choosing, as it is not in man's power. When the man hadn't digged, he wouldn't have found a treasure: here is why, stresses Aristotle, luck is about human matters, while chance about those natural.

Man is in fact the 'principle of actions' [Eth. Nic. 1112b, 31-32]. Things Man can do, 'whose principle is in us', are 'those in our power', and that therefore Man puts voluntarily into achieve (By his exposition Aristotle doesn't distinguish what with the viewpoint of the agent and the one by the philosopher thinking over the action).

On the contrary the principle of involuntary and forced actions is away to the one acting [1110a, 1]:

> It is then generally held that actions are involuntary when done (a) under compulsion or (b) through ignorance; and that (a) an act is compulsory when its origin is from without, being of such a nature that the agent, who is really passive, contributes nothing to it: for example, when he is carried somewhere by stress of weather, or by people who have him in their power. [Eth. Nic. 1110a, 2-4, translation by H. Rackham]

Chance and luck are not causes of what happens out of necessity or at most, are accidental causes. Luck particularly

> regarded as a cause, is the name we give to causation which incidentally inheres to some other end but leading to event we call fortunate. And the significant results of such causes we say 'come by luck'. Thus, since choice implies intention, it follows that luck and intention are concerned with the same field of objects. [Physics, 197a, 5-8, translation by P.H. Wicksteed and F.M. Cornford]

Aristotle emphasize how in the practical activities be necessary, have fortune, or being lucky, the spiritless beings, the animals and the children who can not act can not 'do something by luck', 'because have not the power of choosing'.

Luck matches to human activities as for instance

> For instance, if a man comes to market and there chances on someone he has been wishing to meet but was not expecting to meet there, the reason of his meeting him was that he wanted to go maeketing. [Physics, 196a, 1-5, translation by P.H. Wicksteed and F.M. Cornford]

Human activities, business, are carried out considering in sight of the attainment of a purpose, they are activities with an aim; so the creditor had known his debtor was at the agorà he would have gone there with this purpose to collect his debt.

He went there, but not with such a purpose and only by accident happened to meet his debtor and get back the money, therefore

> it was not due to the fact that mostly, regularly or necessarily, the one who goes [197a] in this place. And the aim, i.e. – receiving the retitution of money – is not among the clear causes in him – in fact it is about causes, resulting from choice and reasoning. In such a case, the one would be told to be prompted by good luck.

> If on the contrary he went there by his own choice and in sight of this purpose – always or most of times he goes there to take money out – then it is not said this 'depends on luck'. [Physics, 196b, 35; 197a, 1-5]

Luck, good or bad as it may be, is not because the execution of the action, a wrong choice, an undesired effect of acting.

However Aristotle emphasizes that luck don't fall in any kind of demonstrative thinking, as thinking be about what is necessarily or for the most part

> On the other side, what depending on chance is not object of demonstrative science. Actually, what depends on chance doesn't present itself neither something necessary, nor something happening for the most part, but it is rather what takes place apart from these two aspects [Sec. An., cap. 30, 87b, 19-20],

anyway luck has to do with deliberative activities of man, at least insofar as it irrupts into accidentally.

Man can only choose on things in his power, even better he is fitted with the 'estimating will deciding on things in his power'. At last the choice is about things to come, from whom 'they take principle'. Now when we deal with social relations, the way it is undoubtedly the case of the meeting between a creditor and a debtor, this possibility is made possible even by the rule on which it is settled.

The choice since which action moves is a sort of calculation (*λογίζεσθαι*) on sight of something, of a purpose, inner to possible things, included, let's say, between two ends: it can not concern things believed impossible or those necessary and endless, neither those passed, nor ultimately those depending on chance, it 'estimates' the 'possible things' (*δυνατά*). Moreover the theory of the 'happy medium' is its explanation.

To go or not to go to the agorà, in Greece 'the real organ of the city' [J. Burckhardt], was in power both of the creditor, and the debtor; both of them, independently one from the other, go to the agorà, where the meeting takes place, accidentally in Aristotle's opinion.

The worry is that the meeting can not be included at all in the intention of the creditor, because it was not his purpose while he was going to the agorà. It was not even the purpose of the debtor, who, as Aristotle says, was at the agorà taking up 'contributions for the party', and not to settle his accounts. The meeting is the outcome of two orders of sequences of facts independent one from the other, that taken separately one from the other have theirs own reasons: both the creditor and the debtor had 'good' reasons to go to the agorà: the meeting was not part of their plans.

The sequence of facts making up the act of going to the agorà by the debtor doesn't affect or has nothing to do with the sequence of facts of the creditor going, in his turn, to the agorà.

Anyway, as we will see, the complementarity of their ratio, debtor/creditor, is what makes peculiarly significant the meeting for both of them.

If we decide to go to the agorà for any business, it is somewhat foreseeable the possibility to meet somebody known, even when this is not the main reason. If I should say: 'Yesterday I went to the agorà to run my business and by chance I met John', my clear saying would become ambiguos to the conversation partner whether I should cancel 'by chance'. My conversation partner as a matter of fact would not understand whether I tell 'an intentional episode or not' [cfr. M. Negrotti, 2007, ivi], i.e. whether it was my intention to meet John or I met him by chance, precisely. Yet this example of Aristotle is suitable so to say only by half. Aristotle, as we have already mentioned, identifies the viewpoint of the agent and the outer observer, is thanks to this identification that he can write 'someone digs a ditch to plant a tree and finds a treasure', and 'get to the agorà by chance and meet whom we were longing to meet, but without having foreseen it'.

If he had taken only the outer observer viewpoint he would have, probably, exposed and discussed the episodes in a different way. In fact, when we keep present what Burckhardt reminds how Aristotle in the *Politikon* aim to distinguish agorà as the place of free men meeting from agorà as place in charge of exchange, business, sales, the agorà comes as place in charge at any rates for meeting among citizens, where, so, it is made possible to anybody to meet anybody else, even whom was not on schedule to be met.

In this case the meeting would be considered inside possible connections belonging to a different order from the one set at the level of the individual agent. By identifying the viewpoint of the agent and of the observer, Aristotle unavoidably adopts the one fo the single agent, aimed by a subjective intention to act,and judges what happens by this point of view. To this side of Aristotle it is possible to add what Descombes tells about the phenomenology concerning the problem of the other, that is to say considers the other as a second I, just like a second first person. [cfr. V. Descombes, 2001, 2004]

This consideration is worth both for the debtor and the creditor, so that for both the meeting sets itself as something irrupting by chance, off rom their intentions. While it is not properly so if we consider the meeting from the viewpoint of the debtor/creditor two complementary positions structure, i.e. in a view that sets together the two positions not considered one apart the other.

At last, the meeting doesn't take place between two individuals whoever they be, but between two individuals bound one to the other by a peculiar and complementary relation, a contract, one has credit with the other, and this is in debt to the first. Burckhardt himself tells us what role had in certain phases of the history of ancient Greece the relation between a creditor and a debtor, such a role that one could enslave the other!

The complementary relation between a debtor and a creditor is therefore a triadic relation, not a dyadic one, one by one. [cfr. V. Descombes, 2006; G. Padovani, 2006]

The meeting between a debtor and a creditor is not an interaction due to the simple composition of independent and lonely acts, that are, in their turn, due to the different intention of individuals that besides are a debtor and a creditor, as this aspect of complementarity makes possible and at the same time regulates their relation.

Furthermore the meeting happens at the agorà, a space, except other things, designed to let people meet, it therefore acts as regulator for the possible relations among individuals, makes them possible since it is by this aspect an open system and not yet closed.

In other words, since themeeting happens in a social space where citizens are made possible to meet, not yet necessarily, but depending on their free will, and it happens among individuals bound by an aspect by a complementary relation, the possibility that an encounter happens 'by chance' is so much probable that becomes awkward talking of chance, if we want to keep to the term the meaning, of something that happens in exceptional circumstances.

At last the writer belives that these phenomena be not explainable as 'the product of Cournot effects, i.e. of the coincidence of causal successions totally independent among them' [R. Boudon, 1990, p. 274] ('le produit d'effets Cournot, c'est-à-dire de la coïncidence de séries causales totalement indépendantes entre elles') [R. Boudon, 1990, p. 268], since the level inside which we consider phenomena, the one of their composition if we like, is the one of totality of possible processes happening when we have just to do with phenomena of composition, of system, whose complexity is in inverse relation to the chance.

By this level, the intention of the subject agent is, so to say, between brackets. This last has undoubtedly a subjective perception of chance, whatever happening and/or happens to him, different from the one possible to be filed by the system level.

To conclude, by the level of the theory of Aristotle and action theories both philosophical and sociological chance gets re-taken in the intentionality of the agent as like as what comes out the expectation and/or from what planned; while by the level of a sociological theory chance gets re-taken in the system of possibilities that social makes possible and partly regulates.

Bibliography

G.E.M. Anscombe, *Intention*, Basil Blackwell, Oxford 1957 (II ed., 1963, 1976, 2000).

Aristotele, *ΠΕΡΙ ΕΡΜΗΝΕΙΑΣ*, It. transl., M. Zanatta (ed.), *Della Interpretazione*, BUR, Rizzoli, Milano 2007.

Aristotele, *Ἠθικὰ Νικομάχεια*, It. transl., C. Natali (ed.) *Etica Nicomachea*, Laterza, Roma-Bari 2003.

Aristotele, *ΦΥΣΙΚΗ ΑΚΡΟΑΣΙΣ*, It. transl., L. Ruggiu (ed.), *Fisica*, Rusconi, Milano 1995.

Aristotele, *The Phisics*, with an English translation by P.H. Wicksteed and F.M. Cornford, Cambridge (Mass.), Havard University Press, London 1980.

Aristotele, *The Nicomachean Ethics*, with an English translation by H. Rackham, Cambridge (Mass.), Harvard University Press, London1982.

J. Burckhardt, *Griechische Kulturgeschichte*, Berlin-Stuttgart 1898-1902.

R. Boudon, *L'art de se persuader des idées douteuses, fragiles ou fausses*, Fayard, Paris 1990.

A.A. Cournot, *Essai sur les fondements de nos connaissances et sur les caractères de la critique philosophique*, L. Hachette, Paris 1851.

V. Descombes, 'Relation intersubjective et relation sociale', in J. Benoist and B. Karsenti (eds.), *Phénoménologie et sociologie*, PUF, Paris 2001.

V. Descombes, *Les institutions du sens*, Les Éditions de Minuit, Paris 1996, It. transl., *Le Istituzioni del Senso*, Marietti 1820, Genova-Milano 2006.

V. Descombes, *Le complément de sujet. Enquête sur le fait d'agir de soi-même*, Gallimard, Paris 2004.

E.E. Evans-Pritchard, *Witchcraft, Oracles and Magic Among the Azande*, Oxford University Press, Oxford 1937, It transl. *Stregoneria, oracoli e magia tra gli Azande,* Angeli, Milano 1975.

C. Lévi-Strauss, *La pensée sauvage*, Librairie Plon, Paris 1962.

C. Natali, *La saggezza di Aristotele*, Bibliopolis, Napoli 1989.

M. Negrotti, 'Il caso vuole', in G. Lanzavecchia and M. Negrotti (eds.), *L'enigma del caso*, Edizioni Goliardiche, Trieste, 2007.

G. Padovani, 'Introduzione. Elementi di una filosofia del sociale', in V. Descombes, op. cit., 2006.

P. Ricœur, *Soi-même comme un autre*, Éditions du Seuil, Paris 1990.

E. Severino, *Legge e caso*, Adelphi, Milano 1979.

Giuseppe Padovani, Dipartimento di Studi Politici e Sociali, Università di Parma

Fumihiko Satofuka

Rationality and randomness in the skill formation of ‘Takumi’ as Japanese traditional culture

Abstract:

Over the centuries, Japan has been able to hold its own in international competition for trade in technologically advanced products, notwithstanding its comparative lack of modern development with respect to other nations. This paper is intended to identify some of the reasons for Japan’s success, and to compare and contrast the Japanese and western approaches towards manufacture and quality control.

Keywords:
logical thinking, rational thinking, intuitive thinking, random thinking, Japanese ‘Takumi’, Western technology, fusion of East and West, right brain and left brain, new technology culture

1 Introduction

The primary concept of recent technological progress is the extent to which computers have penetrated into almost all areas of manufacturing, telecommunications and commodities for which control is necessary. The widespread diffusion of universal control technology can now be seen for the first time.

The computer’s first appearance was in the form of a calculator, and only much later did it become a new control technology. The computer is the first universal control device developed in the history of mankind. The process of computerization has changed not only the features of technology, but also the very face of society; and the transformation continues even today.

From the 1970s to the present day, Japanese companies have succeeded in utilizing the revolutionary capabilities of computers in the development of new products. There have been many suggestions put forward as to why Japan has been able to realize such success: Japanese diligence, special skills, a methodical nature, curiosity, talent for improvement, a high level of education, and so forth, have all been advanced as explanations. However, an important fact regarding Japanese culture has largely been ignored: Japan has a long history of isolation, and there has been a definite pattern in the manner in which she has absorbed influence from foreign civilizations. Without consideration of this factor in relation to the comparatively recent advance of Japanese technology, we should gain only a superficial understanding of the importance of Japan's cultural inheritance. This inheritance is expressed succinctly with the Japanese term 'Takumi', meaning *technique.*

The Japanese tradition of Takumi shows a sharp contrast with the western tradition of technology, based on theoretical thinking and machinery. Figure 1 provides a graphical representation of the relationship between the two approaches. In western countries, logical concepts and scientific knowledge have linguistic and mechanical expression as rationality, while in Japan, intuitively grasped, practical knowledge is expressed in the 'pattern' as randomness. The set of 'patterns', or standardized shapes, is a key concept in Japanese culture. Japanese artisans can easily produce high-quality items by making use of highly developed skills, and many aspects of traditional Japanese craftsmanship are relevant for manufacturing products in Japanese contemporary culture.

Aspects of Takumi make it clear why and how Japan has been able to compete, at a technological level, with more developed countries. In essence, Takumi may be thought of as right-brain technology, whereas western technology may be summarized as left-brain technology. Also, the theoretical modelling for Takumi is made sense of through the history of iron manufacturing, from the Japanese sword in the Middle Ages to the automobile in the postwar period. Therefore I would like, in this paper, to approach the advancement of Japanese technology by focusing on this concept in relation to rationality and randomness in skill formation as Japanese traditional techno-culture.

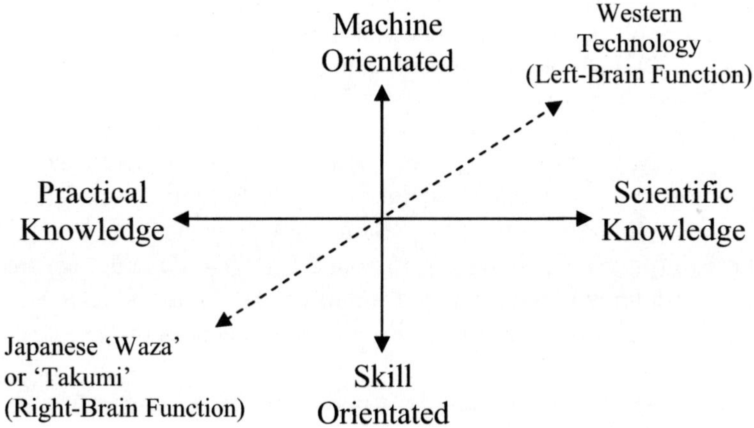

Fig. 1. Relation Between Japanese 'Takumi' and Western Technology

2 Japanese sword, rifle, automobile and emulation

During the history of iron-working in Japan over the past fourteen hundred years, the country has, on at least three occasions, became aware that it had a certain commodity that could dominate international society at a worldwide level. The principal occasions of this kind were offered by the Japanese sword in the Middle Ages (Muromachi shogunate), the Japanese firelock in the Early Modern Age (Nobunaga regime), and the modern automobile in the postwar 20th century.

In the Middle Ages, Japanese swords were exported to China in large quantities, owing to their superiority. Then, in the Early Modern Age, Japan became the world's top-ranked producer of firelocks during the ten years that followed the well-known introduction of Portuguese firelocks to the island of Tanegashima. And as for the automobile, it is needless to say more about it.

A crucial concept regarding international technological competition is that of 'emulation', which can lead to the winning of the competition. The term 'emulation' indicates, in short, the means by which developing countries can win a competition with the advanced countries. The emulation may be expressed in the following formula:

Emulation = Imitation + the competitive condition

That is to say, the developing countries, in order to compete successfully with the more advanced countries, need to hold conditions that the latter do not hold. Without a certain 'something' that the advanced countries do not hold, it would be possible for the developing countries to catch up with the advanced countries, but not to surpass them.

Typical competitive conditions might be cheapness of labour, owing mainly to the payment of low wages, or an abundance of a certain essential raw material. Among the commodities of Japanese origin we can intuitively identify the 'Takumi' as the exterior aspect. Yet it is very difficult to explain the Takumi in words, because of the ambiguity of its content. Even if we can realize the Takumi, we need to explain clearly the reason for the success of its emulation. Therefore, in the next chapter, we shall investigate manufacturing by the Takumi as the origin of Japanese traditional technologies.

3 Manufacturing of steel by 'Takumi'

3.1 Use of the five senses for process management in the Tatara

In 1977, the 100-year-old process of manufacturing steel by means of the Tatara was revived by restoring the ruins of an old Tatara foundry. Abe Yosizou (1902-1995) then became the first 'Murage' (head of the Tatara operation) of the modern age. The Tatara-buki sword-making process is carried out under the leadership of the murage, who uses his five senses and his experience to ensure the production of blades of the finest quality.

From Yosizou's report we can grasp at least the general nature of the process management by the murage in the Tatara. The steel-making operation is one in which the murage is in charge of regulating supplies of charcoal and iron-sand to the furnace, always by judging the conditions inside it. The only measuring devices are the five senses of the murage himself. The process thus constitutes a joint technology system, with the human being making perpetual judgement-based measurements of the process, and regulating it accordingly. In order to manage such a difficult process, Yosizou searched for how he could collect information most usefully. We can show the modelling figure of the Tatara operation.

As the gathering sources for the information we can indicate the following four sources: (1) observation of the flame; (2) observation of the ventilating hole; (3) observation of the 'Noro' (the smelt out impurity from the iron-sand) (4) observation of the sound in the furnace.

3.2 Grasping of the content in sword manufacture

By comparing and contrasting the development of the western sword from the steel plate, on the one hand, with that of the Japanese sword from the steel bowl, on the other, we can try to specify the content of the forging from the viewpoint of the metallurgy, and to contrast the rational forging with the random one.

In modern times, the fundamental qualities of manufacturing materials are homogeneity of quality and a low level of impurities. Western sword manufacture can satisfy all of the necessary conditions rationally, but Japanese sword manufacture can somehow satisfy them randomly.

We can but admire the ability of the Japanese blacksmiths to attain a high level of manufacturing quality through the gathering of information by means of the five senses, without any reliable metallurgical guidance.

3.3 An attempt at the theoretical modelling of manufacture by Takumi

Kazuo Yoshiba, a historian of industrial technology, investigated the manufacture by craftsman (Takumi) in more than 30 kinds of traditional Japanese industries, including an extensive study of traditional bamboo basket-making. He concluded that the manufacture was formed by rules based on human judgement, and that the essence of the manufacture by the craftsman relied on processes of judgement based on the integration of information collected by means of the five senses, all of which serve much the same purpose as that of science in the modern manufacturing processes. Modern manufacturing technology in the post-western industrial revolution can be summarized as the interplay between the following independent four factors: M_1 (Material), M_2 (Machine), M_3 (Method) and M_4 (Man). Such an approach to manufacture is shown in the left-hand part of Figure 2. In contrast, the modelling of the traditional Japanese industry of sword-making by sword-smiths, using the Tatara, is expressed in the right-hand part of Figure 2.

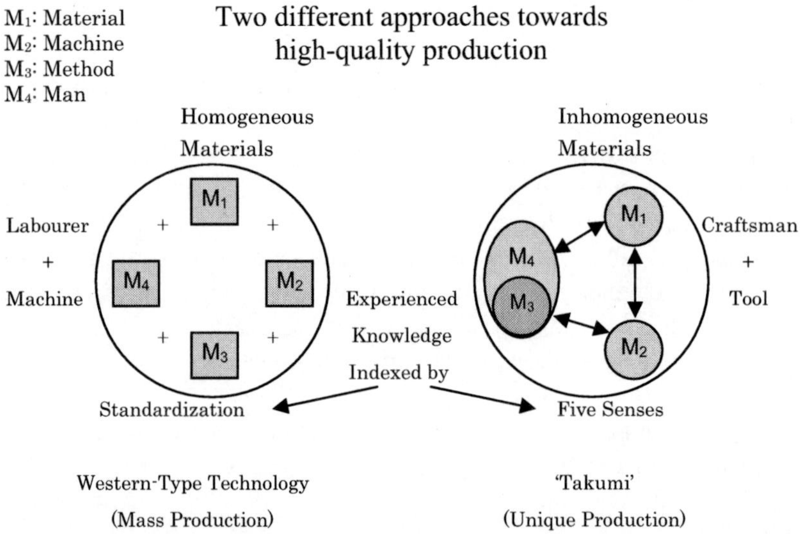

Fig. 2. The Source of 'Takumi'

The illustrating of the two kinds of modelling can more vividly clarify the differences between the manufacturing processes in the traditional industry, on the one hand, and modern mass-production techniques, on the other. Figure 3. provides a comparison between the automobile industries of America and Japan It shows the contrast between Japanese-style Quality Control (QC), which places a greater emphasis on the personal role in ensuring perpetual improvement, and the corresponding western-style QC, which is based more on standardization and homogeneity. We also see here how western-style QC came to 'contaminate' Japanese production, resulting in a sort of hybrid process incorporating elements of both approaches.

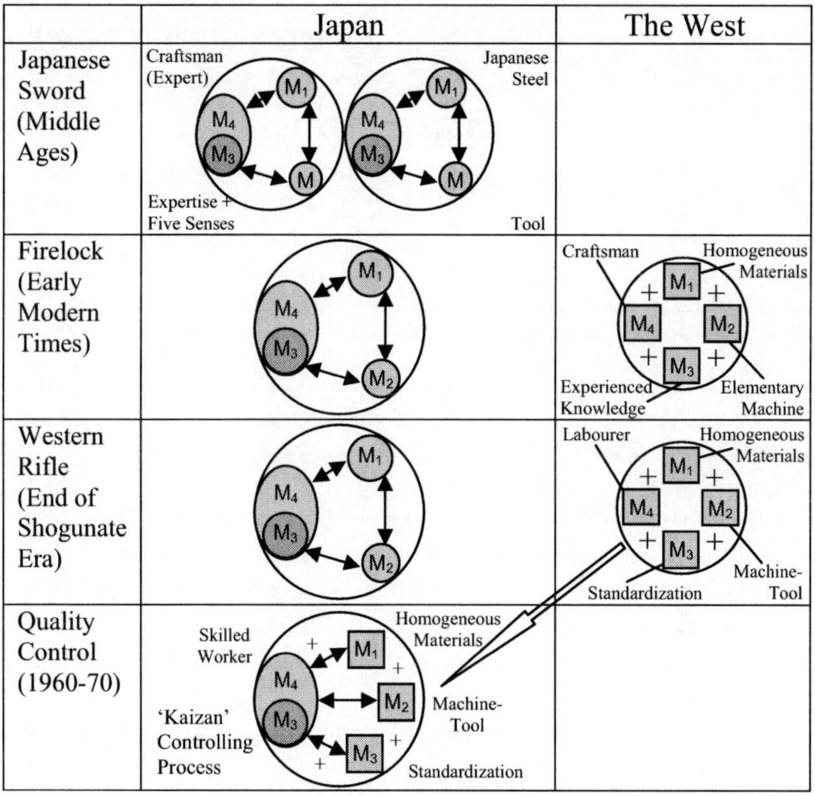

Fig. 3. The Transformation of Japanese 'Takumi'

4 Why the rifle, QC and automobile succeeded to emulate in Japan

Japan became proud of its high rate of steel firelock production – the highest in the world – in the decades following the introduction of the firelock in early modern times. This high rate of production depended in part upon the abundance of wrought iron, and in part upon the abundance of excellent blacksmiths. The firelock, at that time, was a rudimentary device.

In the meantime, the domestication of the western rifles at the end of the Tokugawa shogunate offered a stark contrast. At that time, the rifles which were called for the western gun had a considerable progress as a device, but it had a lot of difficulties for its domestication. As a result a great deal of work machines were imported and domesticated. This kind of rifle production brought about the modernization of mechanical industries.

After the Second World War, the high rate of economic growth brought about the mass production of commercial goods. It reached to the status quo of the automobile production, and became the symbol of Japanese economic prosperity.

In postwar times, Statistical QC (SQC) was introduced from the U.S.A., but it was found to be less than satisfactory for use in Japan. Therefore, in due time, it was transformed into Japanese-style quality control. The American-style SQC was very mathematically orientated, and was mostly carried out off-line by engineers. In Japan it was transformed into the form of QC carried out independently by individual workers on-line.

In my opinion, the Japanese-style manufacturing culture, which was carried out for a long time in unstable conditions, came to be connected to the western-style manufacturing system, which had developed in more stable conditions.

Professor Haruo Shimada compares the Japanese and American automobile industries in his book *The human ware economics*, concluding that the distinguishing feature of Japanese automobile manufacturing is the personal role in the perpetual improvement. This conclusion can be an evidence of the Japanese QC as the basis of a Japanese-style manufacturing system.

5 Conclusion

Japanese 'Takumi' can be expressed in the 'pattern' as randomness, and can easily produce items of high quality by means of highly developed skills, particularly if good materials and tools are available. The above-mentioned aspects of craftsmanship hold for the manufacturing of technologically advanced products, and the main requirements are the same whether you are producing goods by hand or by means of modern industrial processes. The gist of the new technology culture is that it rests on industrial products generated by the fusion of modern science and industrial technology, all based on the foundation of the traditional Japanese work ethic (Takumi).

Bibliography

S. Arima, *The Origin of Firearms and Their Early Transmission*, Yoshi Kaua and Kobun Kan, Tokyo 1962.

I. Inkster & F. Satofuka (eds.), *Culture and Technology in Modern Japan*, I.B. Tauris, 2000.

K. Ishikawa, *The Japanese style quality control*, Nikkagiren, 1981.

JFE 21st Century foundation, *Tatara: Japanese ancient iron making*, 2004.

S. Kobayashi, *A Book Of Colors*, Kodansha International, 1989.

K. Nagata & T. Suzuki, 'The reaction mechanisms and Operation Technique for "Tatara" Steelmaking', in *Iron and Steel*, Vol. 86, No.1, 2000.

T. Ohashi, *Sound and Civilization*, Iwanami Shoten, 2003.

N. Perrin, *Giving Up the Gun: Japan's Reversion to the Sword, 1543-1879*, D. R. Godine, 1979.

S. Sato, 'Japanese Traditional Technologies Using Sand Iron as Iron Resource', in *Ferum*, Vol. 7, No.3, 2002.

H. Shimada, *The human ware economics*, Iwanami Shoten 1988.
J. Suzuki, *The machinery industry in Meiji era*, Minerva Shobo, 1996.

T. Yakushiji, *Techno-hegemony*, Chuko Shinsho 1989.
K. Yoshiba, *Manufacture and Museum*, Ningen No Kagakusha, 2002.

Fumihiko Satofuka, Department of Ecoregion Science, Tokyo University of Agriculture and Technology